Safija Bevandic

THE KEY OF THE UNIVERSE

AUTOR:
SAFIJA BEVANDIC

TITLE:
THE KEY OF THE UNIVERSE

ILLUSTRATOR:
SAFIJA BEVANDIC

EDITOR AND CORRECTOR:
SAFIJA BEVANDIC

PUBLISHER:
SAFIJA BEVANDIC

YEAR OF PUBLICATION AND PRINTING:
2017 – SARAJEVO

PRINT:
IDEA DESIGN – SARAJEVO

CIRCULATION:
200 COPIES

REVIEWER :
ACADEMIC: PROF. DR. MUHAMED FILIPOVIC
CIP – CATALOGING IN PUBLICATION
NATIONAL AND UNIVERSITY OF BOSNIA AND HERZEGOVINA
SARAJEVO 2-17

SAFIJA BEVANDIC
THE KEY OF THE UNIVERSE
[ILLUSTRATOR – SAFIJA BEVANDIC]
SARAJEVO: AUTHOR-2017 - 360 PAGES : ILL 21 CM.
ISBN 978-1-922839-66-4
COBISS – BH – ID 242 60102

TRANSLATED INTO ENGLISH BY :
SAFIJA BEVANDIC

SECOND EDITION and printing in BRISBANE - AUSTRLIJA
BY: INHOUSE PUBLISHING - UNDERWOOD - BRISBANE QLD
ISBN 978-1-922839-66-4
YEAR OF EDITION AND PRINTING 2022

Safija Bevandic

THE KEY OF THE UNIVERSE

Sarajevo, 2017.

PREFACE

My name is SAFIJA BEVANDIC and I am a person who, going through life, I want to get to know life as much as possible and I want to know how to use it, to what extent and when, and when the time is best for use. My goal is to get to know each creation of life, the more that is possible, to reach a conclusion, to decide how to use it, the extent to which it will positively affect my physical and mental health. Since I do not connect emotionally, I conclude by sound logic, which must confirm itself. It was in this direction that I began to explore the Universe and all that is related to the Universe. I wanted to find out the meaning of life on this Earth. I felt there had to be a reason why we were created – I saw that all that was created was created for us and that we have a benefit from this, and that it was created for the benefit of man, and that means we have an important role to play in this galaxy as well as the universe.

I was very interested in this and I started to spend my free time reading books, which I considered to lead to the knowledge of which exists on this earth. All that science has made so far and investigated, and what will be investigated at all, is taken from those books. I saw a great connection between old cultures, always the same information repeats and they all say the same: the universe, peoples and cultures of ancient Egyptians and Mayans in American paintings to the Indians, everything was in the same direction the same calculation; time and light.

My starting point was that someone created all this, someone perfect, because everything created speaks for itself. Everything is perfect, everything should be admired for its perfection. That someone who created it all can be only God, and we can find God in everything created around us and in the word of God. And the word of God is in the books in which we are guided to find everything that we are interested in and every truth.

This truth can be found by people, and only by those who use the power of their mind. I began to take more time to study the word of God, which led me to information which I did not know existed. I studied the books used by the greatest scholars of the world, who have through these books gained the fullest knowledge in mathematics, chemistry, astronomy, physics. On those foundations stands the foundation for all knowledge, and it is found in books that are called the Word of God or the Torah, Zebur , Engil or the Old and New Testaments. In these books, every secret is revealed, for those who with the help of God, want to explore, should be read carefully, because each word has two meanings, every number has its double meaning.

When, with the help of God, I began to discover the secrets revealed in them I decided to write them for people who have no time for such research.

My study lasted thirty years, but from the beginning, from time to time I studied and imagined that I knew enough; that I knew enough to believe that God exists , that I watch how I behave towards God and people. Later, by studying, I noticed that, the same numbers are repeated, such as the numbers: 3, 6, 9 and

12. I noticed this five years ago and started writing down into my notebooks, because these are not just numbers, but through them God wants to reveal an important message to us.
Later, I began to study the word of God more, and since I have often thought about the word of God especially when I lay down, when my thoughts subsided of daily obligations, the truth was revealed to me about those things about which I thought deeply. At first, I thought I would remember what was revealed to me, but unfortunately in the morning, I did not know what information came to me, even though the information was completely clear and understandable to me in the evening I would still forget that information in the morning.
After that, I got up right away and wrote down right the information to not forget it. Afterwards, I learned to always write the information down into my notebooks.
One day, or rather one evening, my eyes opened and I could see more clearly all around me. I got an answer to all my questions that were unanswered until then.
It created a great desire for me to write a part of this understanding of mine. That all those who want to find out who they are, why they are at all born, as for any questions they have which need answers might be answered. Now I can from the information, which is found in this book and in these pictures come up with the most important answer, and that is, that everyone that is born on this earth is born to live forever, regardless that our forefather, the first-born man sinned and therefore lost eternal life, and in which he lived life like in a paradise. Still our heavenly Father

gave for us all, a second chance through Jesus / Isaa / Messiah. Everyone received a second chance, as did Adam / Adem, and he got his second chance, through Christ's / Messiah's death, to recover our lost blessed eternal life. God gave to us all a second chance so we don't say, that because, of Adam / Adem and his sin we have lost eternal life. All of us can lose eternal life because of our own sin. Sin is paid for by Christ / the Messiah's blood, and all who through that blood seek forgiveness of sins, will be forgiven. There is no other name under heaven, nor is there a holy man for us that can forgive our sins except the one who created us , came down from heaven, was born as a man returned to heaven to his father and our father, and who will come to raise us from the dead, and to which we will all stand before him and according to the law, which; he personally gave, to Musa / Moses, on Mount Sinai, will be judged based on our deeds.

When it is said by deeds, because faith without our deeds is dead, we through our deeds we respect God and people, we should do so because it says in the law and in the prophets.

But we cannot earn eternal life with our deeds, for it is written in the word of God, that through one man sin entered into all men it is through Adam / Adem and through his sin that we all inherited sin. It goes on to say that through another man Jesus / Isaa, everyone can be justified from sin and receive eternal life.

In the end, I will state that I would be happy if this information gave someone the opportunity to look at things and events in the way that God wants to show us.

CHAPTER ONE:
REVELATION
THE KEY OF THE UNIVERSE

I feel it's the right name for the content for this book because there is information given to us by the Universe that cannot be changed because they are like seals drawn into our galaxy and therefore which he who created them, determined them to provide us with proof of the truth and confirmation of the truth, so that people have an idea of who they are, where they came from, who created them and where they are going.

Just the name Universe says: this information comes from the space which man cannot control, nor manage that space, for that space belongs only to who created it, to the Lord God, the creator of the Universe as well as all : the visible and the invisible around us. And he did not leave us to guess who we are, where we are from and where we are going after death. And to send messages, and signals to the Universe, that intelligent beings exist and live on this earth

The Universe, sends these signals to us, from the beginning of life on this planet Earth, but people could not understand for the reason that technology and scientific research was not perfected, and people did not have available information about them.

In the times we live in, we have the opportunity to expand our knowledge in all directions of our interest. And among other things, we have a great opportunity to enter into the Universe which is in this time of perfect technology and is accessible to all those who want to study it. If we start from the name itself the Universe, which in the language of simplest understanding means unlimited size and perfection, which means that its perfect and functions harmoniously in space and time. So perfect, that it goes through in the millimetre perfectly. This evidence the Universe itself offers us, and the time has come for all of us to be able to see and read from the universe itself, because some people found and read and they want to share this information with those who have interest in knowledge in that direction, which is vital for each individual.

The first thing we can read and see in the cosmos are the stars and the moon in the sky at night, as well as the sun in daylight hours. The following information is very Important. Everything is round, like our planet earth. New revelations show us that besides these signs or message from space, new characters are coming to complement themselves. And that is, everything created by the Creator, and everything that exists around us, it functions as a circle. That each circle, is an exact circle of 360 °. And that means everything visible and invisible that exists around us works as a perfect 360 ° circle. We will find evidence of this when we look at the rising Sun, as we have a day or a sunset, so we have night. Every 12 hours since creation of life on this earth time goes its own way and towards a certain path and will go that way until its appointed time. The Moon is a circle of 360 degrees, and also goes by its appointed path since the creation of life on this earth and serves us for the night light, as well as for counting days, weeks, and months. Earth is also one 360 ° circle. All the stars are created in a circle, which means, all we see in the sky are 360 ° circles. Altogether, again one large 360 degree circle. Our galaxy is one big circle and also an integral part of the universe of which we have no knowledge of its size. But we have knowledge of everything we need to know. Then on this planet of ours, everything functions as a circle of 360 ° from the first moment that time began to count. It is written in the first book of Musa / Moses or the book called ,, GENESIS'' on the first page.

Picture 1 (THE DARKNESS OVER THE ABYSS)

In this picture we see the earth that existed in the Universe in which there was no life. It was dark and it was only one round of 360 °, right up until God created life on it.

Verse 1. "In the beginning God created heaven and earth."

Verse 2. "The earth was without form and desolate, and there was darkness over abyss; and the spirit of God is lifted up over the water."

Verse 3. "And God said, 'Let there be light!' And there was light. "

Verse 4. "And God saw the light that it was good; and God separates light from darkness."

Verse 5. "And God called the light Day, and the darkness he called Night. And the evening and the morning were the first day."

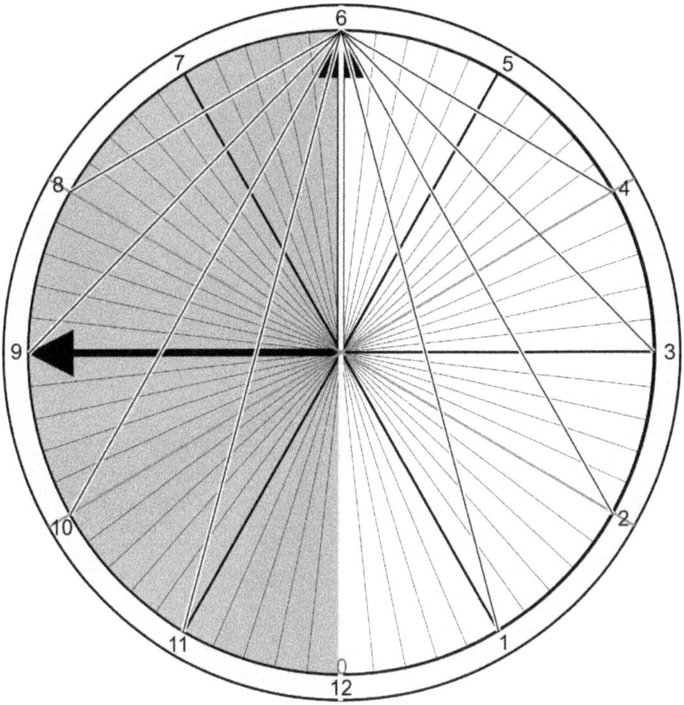

Picture 2 (BEGINNING – BEGINNINGS)

Picture 2 shows the light which separated the darkness in which it began to count the time. It is clear that the Creator immediately began to count the time from the beginning. From the very beginning of life on earth, from the first second, the time started immediately.

Every day begins with the " And it would be evening, And it would be morning." That means the day is starting the moment the Sun goes down, which means first Night then Day. Then comes the order of creation, in the same book of creation of life.

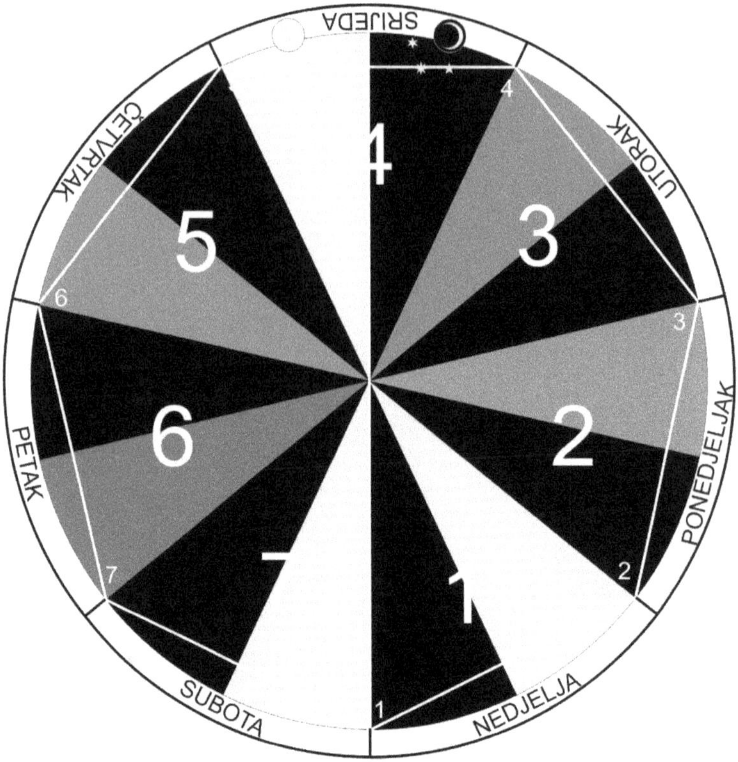

Picture 3 (DAYS OF CREATION).
1.(NEDELJA – SUNDAY); 2.(PONEDELJAK – MONDAY); 3.(UTORAK – TUESDAY); 4.(SRIJEDA – WEDNESDAY); 5.(CETVRTAK – THURSDAY); 6.(PETAK – FRIDAY); 7.(SUBOTA – SATURDAY)

Verse 6. "And God said, Let there be a firmament in the midst of the waters, and let it divide the waters from the waters."

Verse 7. "And God made the firmament, and divided the waters which were under the firmament from the water which were above the firmament: and it was so."

Verse 8. "And God called the firmament Heaven . And the evening and the morning were the second day."

Verse 9. "Then God said, Let the waters under the heaven be gathered together unto one place, and let dry land appear: And it was so."
Verse 10. "And God called the dry land Earth; and the gathering together of the waters called the he Seas: and God saw that it was good."
Verse 11. "Again God said, Let the Earth bring forth grass, the herb yielding seed and the fruit tree yielding fruit after his kind, whose seed is in itself, upon the earth: and it was so."
Verse 12. "And the Earth brought forth the grass and herb yielding seed after his kind, and the tree yielding fruit, whose seed was in itself, after his kind: and God saw that is was good."
Verse 13. "And the evening, and the morning were the third day."
Verse 14. "And God said, Let there be lights in the firmament of the heaven to divide the day from the night; and let them be for signs, and for seasons, and for days, and years."
Verse 15. "And let them be for lights in the firmament of the heaven to give light upon the earth: and it was so."
Verse 16. "And God made two great lights; the greater light to rule the day, and the lesser light to rule the night: he made the stars also."
Verse 17. "And God set them in the firmament of the heaven to give light upon the Earth."
Verse 18. "And to rule over the Day and over the Night, and to divide the light and to divide the light from the darkness: and God saw that it was good."

Verse 19. "And the evening, and the morning were the fourth day."
Verse 20. "And God said, Let the waters bring forth abundantly the moving creature that hath life, and fowl that may fly above the Earth in the open firmament of heaven."
Verse 21. "And God created great whales and every living creature that moveth, which the waters brought forth abundantly, after their kind, and every winged fowl after his kind: and God saw that it was good."
Verse 22. "And God blessed them, saying. Be fruitful , and multiply, and fill the waters in the seas, and let fowl multiply in the Earth."
Verse 23. "And the evening, and the morning – were the fifth day."
Verse 24. "And God said, Let the Earth bring forth the living creature after his kind, cattle, and creeping thing, and beast of the Earth after his kind: and it was so."
Verse 25. "And God created the beast, after his kind, and cattle after their kind, and everything that creepeth upon the earth after his kind: and God saw that it was good."
Verse 26. "And God said, Let us make Man in our image, after our likeness: and let them have dominion over the fish of the sea, and over the fowl of the air, and over the cattle, and over all the Earth, and over every creeping thing that creepeth upon the earth."
Verse 27. "So God created man in his own image, in the image of God created he him: male and female created he them."
Verse 28. "And God blessed them, and God said unto them, Be fruitful, and multiply, replenish the Earth, and subdue it: and have dominion over the fish of the sea, and over the fowl of the

air, and over every living thing that moveth upon the earth."
Verse 29. "And God said Behold, I have given you every herb bearing seed, which is upon the face of all the earth, and every tree, in the which is the fruit of a tree yielding seed; to you it shall be for meat (food)."
Verse 30. "And to every beast of the earth, and to every fowl of the air, and to everything that moves on the earth, wherein there is life. I have given every green herb for meat (food): and it was so."
Verse 31. "And God saw everything that he had made, and, behold, it was very good. And the evening and the morning, were the sixth day."

CHAPTER TWO

Thus the Heavens and the Earth, were finished, and all the host, of them.
Verse 2. "And on the seventh day God ended his work which he had made: and he rested on the seventh day from all his work which he had made."
Verse 3. "And God blessed the seventh day, and sanctified it: because that in it he had rested from all his work which God created and made."
Verse 4. "These are the generations of the heavens and of the Earth when they were created, in the day that the LORD GOD made the Earth and the Heavens."
You can see the 360 ° circle, this is also the first circle in which the Creator covered the first week he created life on this planet.

In that first week, God gave us a guide and charted the direction of everything what we will go through, all the events that will be important of significance for our future life. The Creator also gave confirmation of the Universe in that first week, so that people would not guess what is true and whether exactly, it is true. All this, as evidence, is found in the picture of the first week, as well as in the events described.

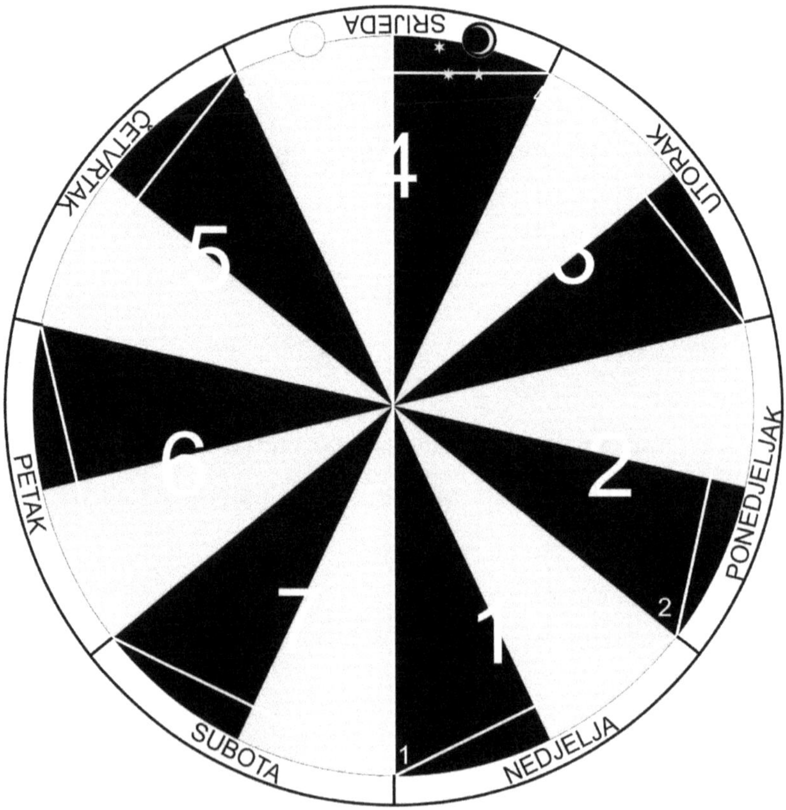

Picture 4 (FIRST - WEEK).
NEDEJA – SUNDAY; PONEDELJAK – MONDAY; UTORAK – TUESDAY; SRIJEDA – WEDNESDAY; CETVRTAK – THURSDAY; PETAK – FRIDAY; SUBOTA - SATURDAY

In the book of Genesis or the first book of Musa /Moses/ p.1.
In this picture we see a circle from the first week which is divided into seven equal parts .There are seven days in one week. Because of this, the name is seven parts, or seven days.
From the first moment of creation, in which God created light, he created the first second in which he started counting time. He continued with the creation of the first day, where it says, in the above verses, "And it would be evening, and it would be morning ; the first day."
That means, firstly it will be evening, then morning and he gave us the information that each day is divided by half, into evening and morning, to darkness and light. In that first week he showed us: time, space and himself. So everyone who seeks him will not be lost, but will find eternal life.
Now, let's continue with the week and the second from the first second until the time and first night. From day one time started to count for this planet. However since time through history and our lives was totally changed, we now have a different calculation of business-time or world time and it's no longer as the time that was created by the creator. There was a shift in hours, days and years in our calendars so it would be easier for world-wide communication for all people and it could be used for economic purposes and which makes it more accessible. To make communication easier in the exchange of business ideas and easier cooperation in the economic needs of society.
But one important fact is that everything God created was not

created to go in that direction, because the Creator gave us all these signs so that we would go in his direction, in the direction of truth and the path that leads us to the place where he created us. And that is the Garden of Eden or (Janet), who the first man, Adam / Adem lost.

The beginning of time and day begins in this order:

The first day of the week is Sunday;

The second day of the week is Monday;

The third day of the week is Tuesday;

The fourth day of the week is Wednesday. Wednesday is the day that shows us that it was also the middle of the week, and that name was given by whoever created all the days and gave them names too. All that God created and said has great importance, because he did not say a word in vain so Wednesday has great importance;

The fifth day of the week is Thursday;

The sixth day of the week is Friday;

The seventh day of the week is Saturday. The seventh day, which is also the seventh part of this 360 ° circle indicates directly that this circle is closed with 7th part or with 7th day - Saturday. Saturday connects the first day - Sunday and the sixth day - Friday. It stands between them and sends us information of great importance, such as an indication of the direction of the movement of time or that it is connected between six weekdays and one non-working day of the week.

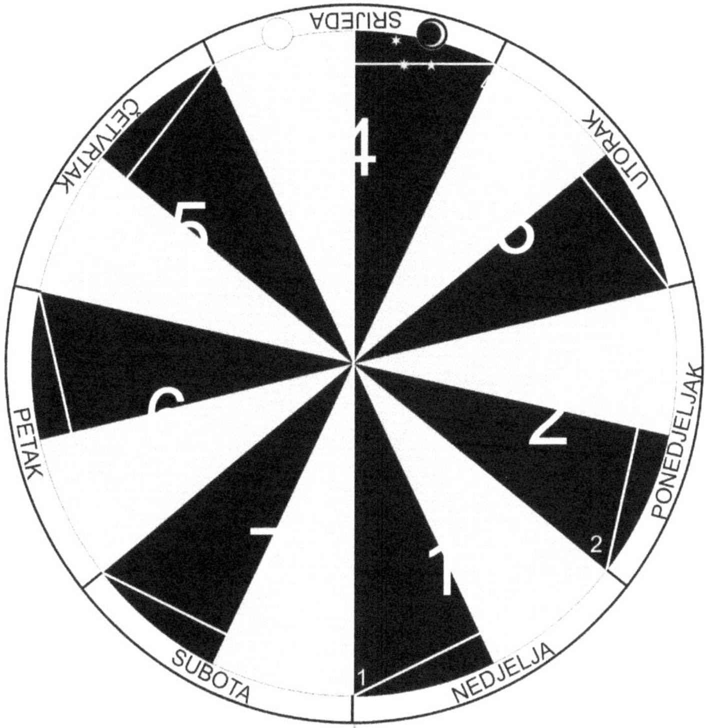

Picture 4a. (FIRST WEEK and FIRST ALFA AND OMEGA).

This figure shows that the seven parts make up one 360 ° whole or seven days a week and to this day they are still being counted and named. In some parts of the world, the first day of the week is called Sunday, and that would be translated: does not work. However, in many countries of the world, in spoken English languages, as well as German and other languages, Sunday is called sunny or the Sun day, or a sunny day. Since the Creator created the light in the first second of creation, and that was in the first day of the creation, is the day of light at the first day of the week, and it suits its name.

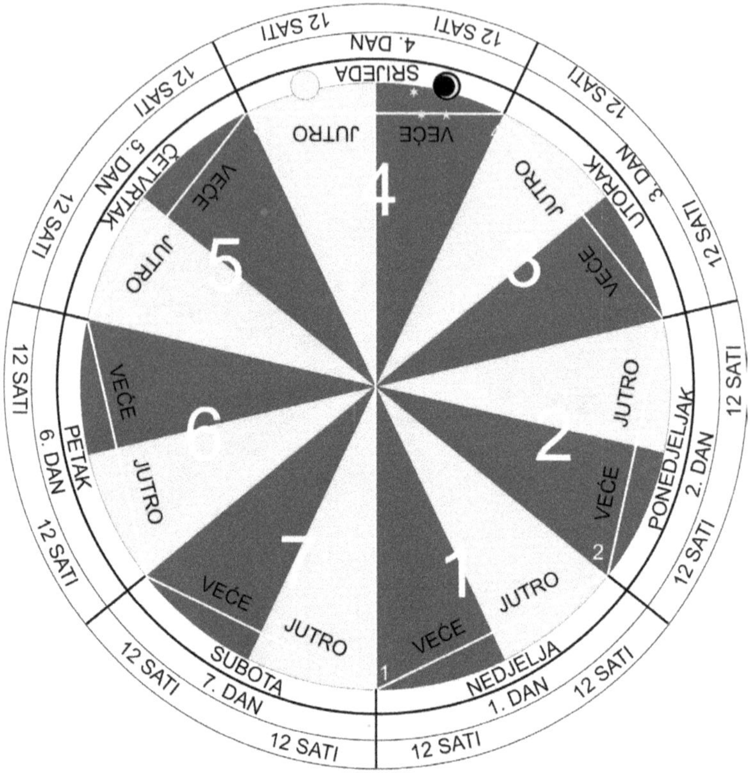

Picture 5 (AND IT WAS EVENING AND IT WAS MORNING)
VECE – NIGHT; JUTRO - MORNING

In this picture, we see each day split in half. The way it was created by the Creator: "And it would be evening, and it would be morning - the first day." Every day has a night and day. Since the night comes first, so is the first half day night, the other half day is light or day. The Creator said, "And it' would be evening, and it would be morning," And this means that the day, and its time is counted at twelve hours. At twelve hours in the night, God is not creating anything and; and everything he created He created in daylight, per day.

If we go from the first second to "And it would be Evening", we see that night separated from the day and so it will be as long as there is life on this earth. But as we move on to other days of the week and it comes Monday, it's the second day of the week. After Monday comes Tuesday, as the third day of the week.

But on the fourth day of the week of life on this Earth God created the Sun, the Moon, and all the host of Heaven. This information indicates that, until the fourth day there was neither the Sun nor the Moon and that the Creator (created) it in the middle of the week until Wednesday in his light. God is light and all things he created in those hours and days, it is forever under God's control, and that people cannot, control the creations of God. And the light that God created in the first second passed through all the Earth and divided the darkness from light. And the direction of light was until Wednesday of the fourth day of creation which fulfilled as the middle of the week. That's where the Creator confirmed that Wednesday is exactly the middle of the week. On Wednesday in the middle of the week God's light meets the light that God created for us and that is the Sun, which receives the light of God , and remain according to God's will, to shine for us, and to give life on this earth. The next day, behind Wednesday, is Thursday, and then Friday. Friday is the sixth day of the week and in that sixth day God created Man, as the CROWN of creation. Thus it is finished, Heaven and Earth and all the host of Heaven. It shows us that the Creator created everything in six working days.

And by the seventh day God finished his works of creation.

And on seventh day, God ended his work which he had made ; and he rested on the seventh day from all his work which he had made. And God blessed the seventh day, and sanctified it, because in it he had rested from all his work which God created and made. These are the generations of the heavens and of the Earth when they were created, in the day that the Lord God made the Earth and the Heavens.

From these verses taken from the book “Genesis “, from the first book of Musa / Moses; Head 2;

verse 1. 2. 3. and 4. Here the Creator showed us that the seventh day is the Sabbath, for we have seen that the Sabbath fits between Friday and Sunday, and connects Friday with Sunday, Sunday as the first day God created, and the sixth day Friday, which he ended making Life on this earth; The Sabbath, being the seventh day in creation and God has separated it from these six working days. And every week, since the beginning from the first second since time started to count, it has six working days and one holy day, In these six working days we have 6 x 12h nights and 6 x 12h daylight, it is 72 hour nights and 72 hour daylight. But we also have a Sabbath, and she separates holy (12 hour Night and 12 hour daylight) from 144 working hours.

When we add up that first week we will get 144 working hours, and 24 holy hours.

From this result, we also see that the Creator separated hours and days and it has a very important meaning, it will lead us to new information that the Creator sends to us and we need to know in order to see what the Universe tells us through time. And what that means over time, we will see exactly through time. Moving on, from the first week, keeping track of time and passing through time we see new facts and that is that time began to count starting from the first second of the original time. That first second always brings us back to reality and to the path of truth.

HOURS OF THE FIRST-WEEK

When we start to add up weeks, from the beginning of becoming, we will get next the result:

One week has 144 working hours.

Ten weeks has 1.440 working hours.

One hundred weeks have 14.000 working hours.

The Thousand Weeks has 144.000 working hours. Since the number 144.000 has of particular importance in the calculation of time, that number will be here also specially recorded.

This was the working time of thousands of weeks since life was created on Earth.

Now there is time, left from the holy hours.

The first week from the beginning of time and life on the Earth: has 24 holy hours:

For ten weeks there are 240 holy hours.

One hundred weeks have 2.400 holy hours.

There are 24.000 holy hours in a thousand weeks.

That's the correct number for a thousand weeks. Altogether, holy and working hours amounts:

144.000 + 24.000 = 168.000 hours in one week:

Or 10.000 weeks has 1.440.000 hours:

Or 10.000 weeks has 240.000 holy hours.

We will stop at this place for hours and move on to the days.

Why is time counted by weeks? The Creator gave us that instruction and showed to us that in the weeks, we will find the direction of time - that we need to know about it.

Lord God, when he created life on this planet, he created it in one week, in six working days, from Sunday to Friday.

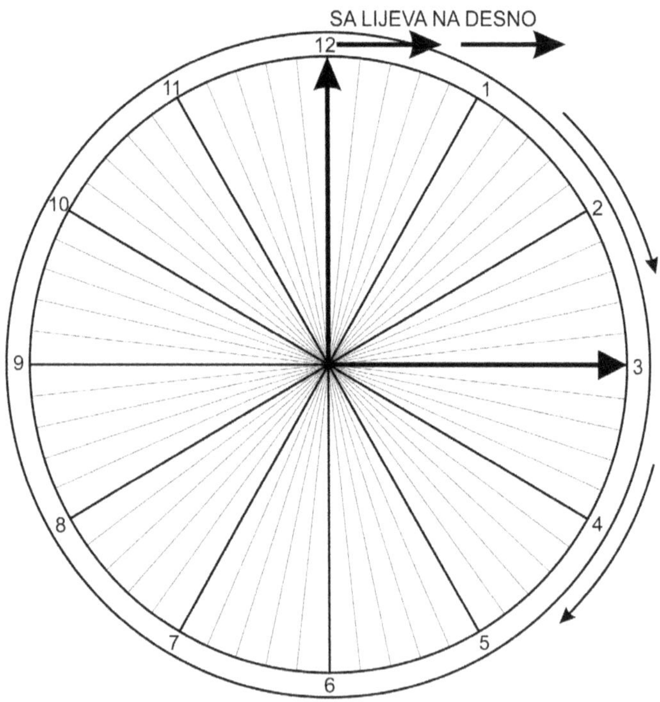

Picture 6 (AN HOUR OF TIME) (FROM LEFT TO RIGHT)

THE HOURS OF TIME:

This picture shows us the clock we put on our hand every day or have in our own house. And just this simple watch is a roadmap to take us to where it should take us, because it was ordered by him by the Creator.

The clock is also a 360 ° circle divided into 12 parts and it also starts to count down from the first second.

On it, every day, we watch as time goes by. An hour to us is one big daily warning. In it, and through it we see how we go through time as well.

We all have our time that we need to take advantage for the future time in which, for sure one day we will all depart from. All those who don't have time to stop to see that time is one of the most important things on this earth, and to use their time, which they have been given to use in good thinking and in the correct use of that time they will lose the battle the Creator gave them and they will enter a time where there will be no more time and where we will not count time anymore because on this Earth everything is limited by time. And everything has its own shelf life.

In addition, time gives us information on where we are currently as well as the time that has passed and time that will come - we can read about it through the clock we use in our daily lives. But the clock also changed appearance due to the needs of society, and on it we see time going in the opposite direction from the time that was set by the one who created the time. Our watch shows the time that goes from left to right and is called the left clock.

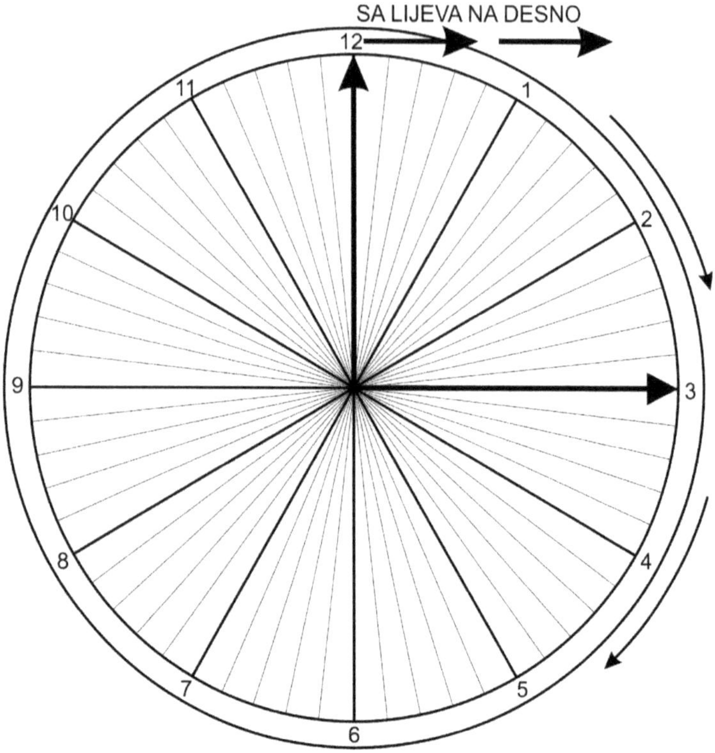

Picture 6a (AN HOUR TIME LEFT to RIGHT).

It shows time from left to right that leads us to the wrong one direction of calculating time.

The Creator set the time in the direction from right to left, just like he also set the days of the week in the direction from right to left, and that it's called the right clock.

This is very important information because the left-to-right direction leads us in the opposite direction from the truth and from God and his direction is to the left, goes into darkness with no way out.

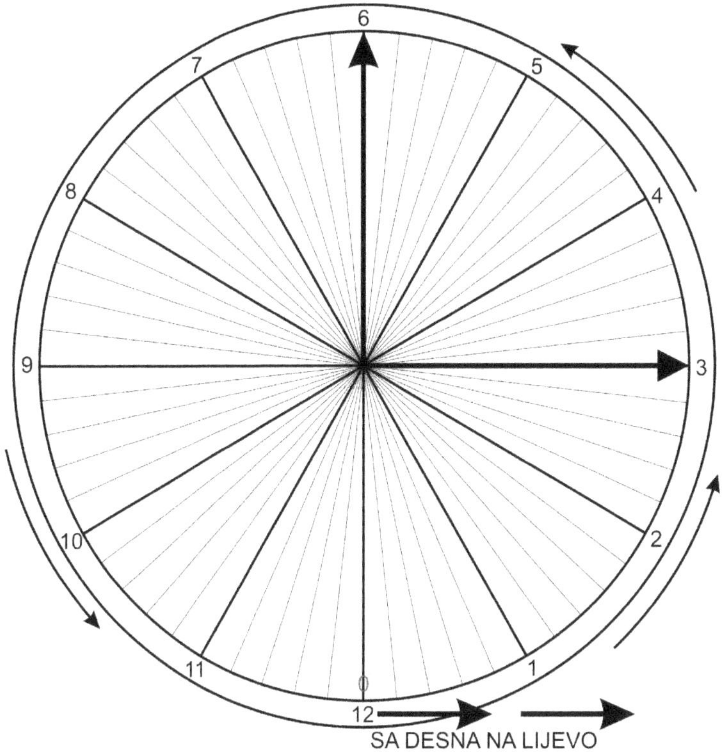

Picture 7 (AN HOUR OF TIME FROM RIGHT TO LEFT)

Shows the other right-to-left direction, its direction is to the right, and which enters the Milky Way. It is a path that leads to a different galaxy and is the only path that has an exit from this galaxy. It is very important information to know.

Time that shows the clock, but clock turned clockwise from right to left , we need to monitor that time to get information for which we did not know existed. That is the direction into time that is everlasting.

Here we see a 360 ° circle divided into 12 parts.

Everything that has a 360 ° circle has the perfect creation and the seal called Omega, but every Omega must have an Alpha which means that the beginning merges with the ending in the thousand parts in the millimetre.

In simple language, something is continuously being restored and it does not have a border. That means, this watch has no end, but it was the beginning of that first second. Seconds are constantly repeated like minutes, hours, days, weeks, as well months as well as years as well as thousands of years. All this will happen again like Alpha and Omega until the time when ones time is up which the Creator had specifically assigned to him.

That means, the watch we have on our hand or in our everyday houses are very important because we see all the information which is connected to life on this Earth. Both for the present, and for the life that we are going to after this life. On those who care to understand this will be able to understand it because they will take

Own time in honest reseach, written explanation.

their time in honestly exploring what is written.

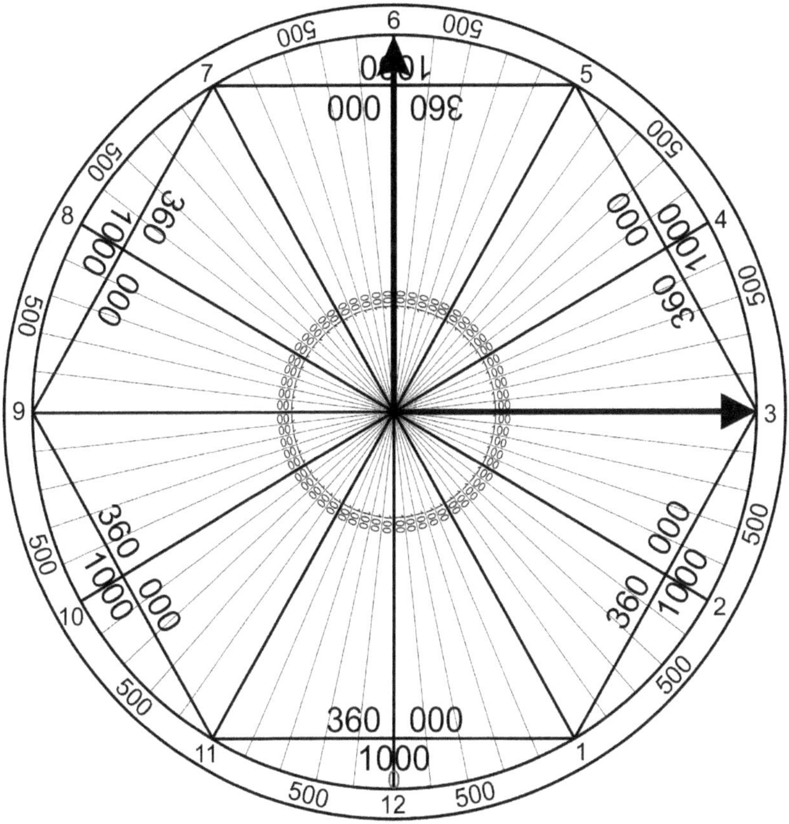

Picture 8 (TIME IN TIME)

We see the time from the first second, the continued time, and the time until 6000 years.

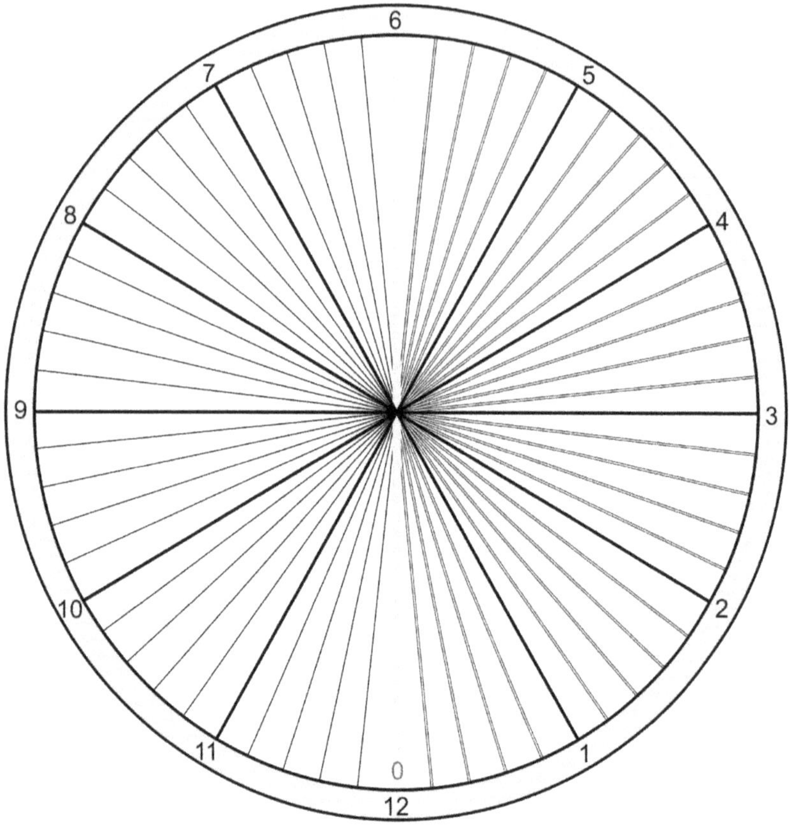

Picture 9 (FIRST SECONDS).

In these pictures, at the same moment, we see seconds, minutes, hours, weeks, months, years, hundreds of years, thousands of years.

In this hour of time, to see all this, we must first split up seconds. The Second is the first creation of time in which we live on this earth by, a second turns a minute into a moment when it ticks into 60 seconds.

One minute has 60 seconds.

One hour has 60 minutes.

One half day has 6 hours or 6 x 60 minutes, which is equal to the sum of 360 minutes.

This line is where the Creator divided the light from darkness and on which the first second arose.

Then, this place is exactly halfway through the year.

This place is also halfway through the week.

In this place the Creator created the Sun and the light that continues to sustain life on this Earth.

This place is half of the 6.000 years that makes Omega whole.

What does the whole Omega mean? That means that this is the place where Omega passed through Alpha and that is where it made the beginning and end, like light dispels darkness and creates a day or separates truth from lies and life from death.

So Alpha shows, through time, seconds, minutes, days, years, and all we need to know,

which in that place composed Omega and Alpha, which will show us the following information and pictures.

So Alpha shows, through time, seconds, minutes, days, years, and all that we need to know, and that is exactly in this place that Omega and Alpha combined. This will be shown in the following information and pictures.

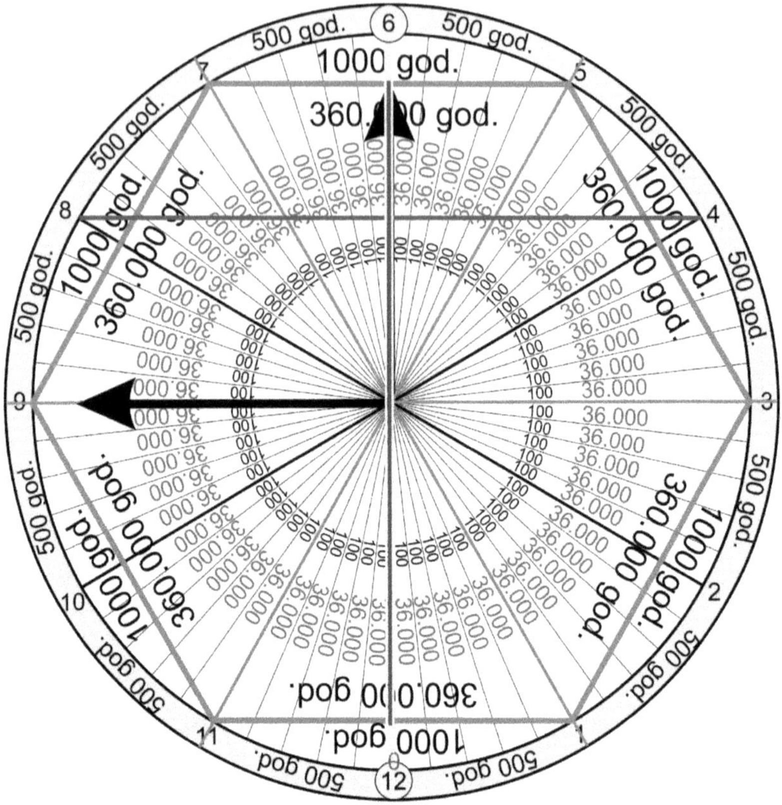

Picture 10 (TIME AND SPACE THROUGH NUMBERS 3 AND 6)

We see an hour. Which instead of 1 minute show us 100 years, and instead of 5 minutes shows us 500 years and instead us 10 minutes shows us 1000 years.

We see the middle painted in yellow, and it shows us the direction of movement, light from the first second to the half of the one week,(Wednesday) or up to half the time of one hour, or half the time of one day, or half the time of one year or to half the time of 6.000 years.

Here we see is an hour of time, which shows us 100 years instead of a minute as well as the first second of creation when it was created, so we can see days and years that have already passed, as well as years that will come.

If we start from the first year and it has 360 days, then :

10 years will have 3.600 days:

100 years will have 36.000 days:

1000 years will have 360.000 days:

And that means:

1.000 years has 360.000 days:

2.000 years has 720.000 days:

3.000 years has 1.080.000 days:

4.000 years has 1.440.000 days:

5.000 years has 1.800.000 days:

6.000 years has 2.160.000 days:

Since six days a week is counted in working hours, for in six days God created all things, and on the seventh day, he rested from all of his works, and dedicated the seventh day and separate it from the working days, so we see that the seventh day counts especially from the moment of creation and through all this 6.000 years has special significance.

The word of God says, and this hour of time confirms, that God created man on the sixth day on Friday at the 144th hour. The first week of creation.

In the book of Moses:

5. Head: Verse 1

"This is the generations of Adam. In the day that God created

man, in the likeness of God him."
Verse 2. " Male and female created he them; and blessed them, and called their name Adam / Adem in the day when they were created."
Verse 3. "And Adam / Adem lived 130 years, and begat a son in his own likeness, after his image; and called his name Seth."
Verse 4. "And the days of Adam after he had begotten Seth were 800 years: and he begat sons and daughters."
Verse 5. "And all the days that Adam / Adem lived were only 930 years: and he died."
Verse 6. "And Seth lived 150 years, and begat Enos."
Verse 7 "And Seth lived after he begat Enos 807 years, and begat sons and daughters".
Verse 8. "And all the days of Seth were 912 years , and he died."
Verse 9. "And Enos lived 90 years and begat Cainan."
Verse 10. "And, Enos lived after he begat Cainan for 815 years, begat sons and daughters."
Verse 11. "And all the days of Enos were 905 years; and he died."
Verse 12. "And Cainan lived 70 years, and begat Mahalaleel."
Verse 13. "And, Cainan lived after Mahlaleehl 804 years, and begat sons and daughters."
Verse 14. "And all the days of Cainan were 910 years; and he died."
Verse 15. "And Mahalaleel lived to be 65 and begat Jared."
Verse 16. "And, Mahalaleel lived after he begat Jared or 830 years, begat sons and daughters."
Verse 17. "And all the days of Mahalaleel were 895 years; and he died."

Verse 18. “And Jared lived an 162 years, and he begat Enoch.”
Verse 19.“And Jared lived after he begat Enoch: 800 years, and begat sons and daughters.”
Verse 20. “And all the days of Jared were 962 years: and he died.”
Verse 21. “And Enoch lived, 65 years, and begot Methuselah.”
Verse 22. “And Enoch walked with God after he begat Methuselah 300 years, and begat sons and daughters.”
Verse 23. “And all the days of Enoch were 365 years.”
Verse 24. “And Enoch walked with God: and he was not; for God, took him.”
Verse 25. “And Methuselah lived 187 years, and begat Lamech.”
Verse 26. “And Methuselah lived after he begat Lamech 782 years, and begat sons and daughters.”
Verse 27. “And all the days of Methuselah were 969 years; and he died.”
Verse 28. “And Lamech lived 182 years, and begat a son.”
Verse 29. “And he called his name Noah, saying, “This same shall comfort us concerning our work, and toil of our hands, because of the ground which the LORD hath cursed.”
Verse 30. “And Lamech lived after he begat Noah 595 years, and begat sons and daughters.”
Verse 31. “And all the days of Lamech were 777 years; and he died.”
Verse 32. “And Noah was 500 years old, and Noah begat Shem, Ham, and Japheth.”
In this same book, “Existence” or 1 Musa / Moses Book; chapter 6; (verse 5).

"And the Lord, saw that the sin of man is great on earth, that all thoughts and hearts are always evil."

Verse 6. "And it repented the LORD that he had made man on earth and it grieved him at his heart."

Verse 7. "And the LORD said, I will destroy man whom I have created from the face of the earth ; both man, and beast, and the creeping thing, and the fowls of the air; for it repented me that I have made them."

And God said:

Verse 17. "And, behold, I, even I, do bring a flood of waters upon the earth, to destroy all flesh wherein is the breath of life, from under heaven; and everything that is in the earth shall die."

Verse 18. "But with thee will I establish my covenant; and thou shalt come into the ark, thou, and thy sons, and thy wife, and thy sons 'wives with thee."

And God told Noah / Nuhh to build a ship into which he, his wife, his sons, and their wives, take refuge from the flood including all living things that God planned to survive the flood, by their kind and by their number. There would only be eight human souls. And of them the earth was multiplied."

Verse 22. "And Noah did as God commanded him to do, so he did."

CHAPTER 7

Verse 6. "And Noah was 600 years old when the flood of waters was upon the earth."

Verse 11. "When Noah was 600, that year of the second month, On the 17th day ,of the month the same day were all the fountains of the great deep broken up, and the windows of

heaven were opened."
Verse 12. "And it rained on the earth for 40 days and 40 nights."
Verse 18. "And the waters prevailed, and were increased greatly upon the earth; and the ark went upon the face of the waters."
Verse 20. "Fifteen cubits of water went over the hills."
Verse 23. "And all flesh was destroyed, on the earth, both men and livestock, and whatever ate, and the birds of the sky, everything I say, is exterminated from the earth, only Noah / Nuhh, the rest and whatever was in the ark with them."

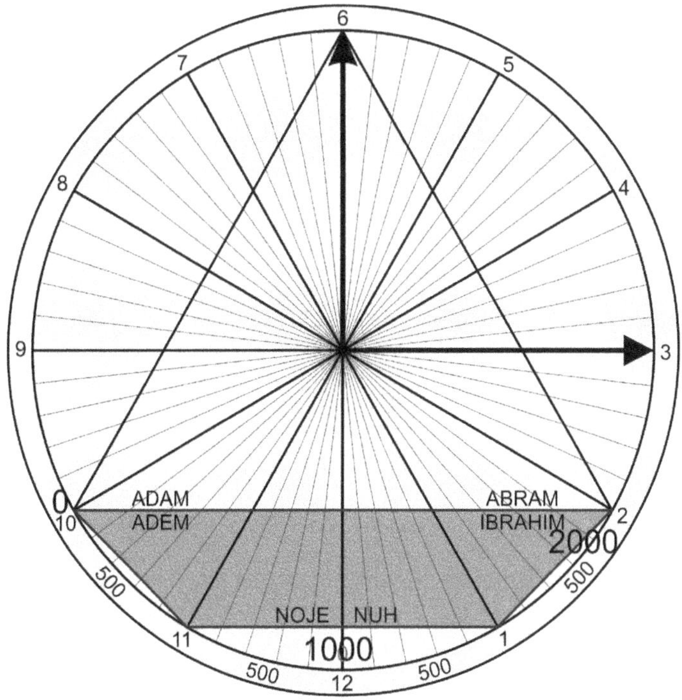

Picture 11 (FLOOD).

It shows the accuracy of these events within an hour.

In this picture of time we see the exact events that happened and which the Universe shows and confirms. Exactly 1.000 years after Adam / Adem lived as a sinner on this earth Noah / Nuhh was born.

Noah survived the flood with his family, only eight souls and what God sent into the ark survived only.

At this hour the time that has passed and which is confirmed through the universe sees the image of the ark during this period and in those years in which was going through these big events, where people lived very long and sinned a great deal. All the blessings they received from God, they have used against God, satisfying their sinful desires and enjoying in what was forbidden by God, so God he repented that he had made man, and the penalty for sin was death, and they received it as their salary. Moral darkness has never been as dense as in that one period.

The Book of Moses:

Chapter 7:

Verse 17. "And the flood was forty days upon the earth; and the waters increased, and bare up the ark, and it was lift up above the earth. And the waters prevailed, and were increased greatly upon the Earth; and the ark went upon the face of the waters. And the waters prevailed exceedingly upon the Earth; and all the high hills, that were under the whole heaven, were covered."

Fifteen cubits reached the water above the hills, as it covered every moving body on Earth, birds and cattle, and beast, and all that moved on the Earth, and all Men. Everything that had a soul,

everything that which was on dry land had died. And everybody alive on Earth was destroyed, and Men and cattle, and the birds of the sky, all, was utterly destroyed from the earth, only Noah / Nuhh was left and what was with him in the ark.
The fountains also of the deep and the windows of heaven were stopped, and the rain from heaven was restrained;
And the waters returned from off the Earth continually: and after the end of the 150 days the waters were abated.
And the ark rested in the seventh month, on the 17th day, of the month, upon the mountains of Ararat.
And the waters decreased continually until the tenth month: in the tenth month, on the 1st day of the month, were the tops of the mountains seen.
601 AD, century Noah / Nuhh, 1st day of the first month the waters were dried up from off the earth:
And in the second month on the 27th day of the month, was the earth dried.
And Noah / Nuhh and his sons went out, and his wife, and his sons wives with them. All the beasts, all the little animals, all the birds, and everything moving around the Earth by its kind, it came out of the ark.
And the Noah / Nuhh made an altar unto the Lord, and took every pure cattle and of all the birds that are pure , and he brought sacrifices upon the altar.
And God blessed Noah / Nuhh and his sons and said to them, "Rejoice and multiply, and fill the Earth, and all the beasts of the

Earth, and all the birds within the heavens, and all that moves upon the Earth, and all the fishes of the sea, let them be scared by you. Everything is placed into your hands. Whatever moves and lives, let it be for your eating, I give it all to you just like the green grass. But flesh with the life thereof, which is the blood thereof, shall ye not eat.

And surely your blood of your lives will I require; at the hand of every beast will I require it, and at the hand of man; at the hand of every man's brother will I require the life of man.

Who sheds Man's blood, by man shall his blood be shed: for in the image of God made he created man.

And God said to Noah / Nuhh and his sons,

"I make my vow to everyone, and from now on no flesh shall not be destroyed by the flood, neither shall there be more floods to destroy the earth."

And God said, "Here is the sign of the covenant I make between myself and you. I have cast my rainbow into the clouds to be a sign of a covenant between me and you. So when I bring the clouds down to Earth, you will see a rainbow in the clouds, and you will remember my vow that is between me and you, and every soul that lives live in a human body, and there shall be no more floods to destroy anybody anymore.

The rainbow will be in the clouds for a long time so I'll look at it and be reminded of the covenant between God and every soul which dwells in everybody on Earth."

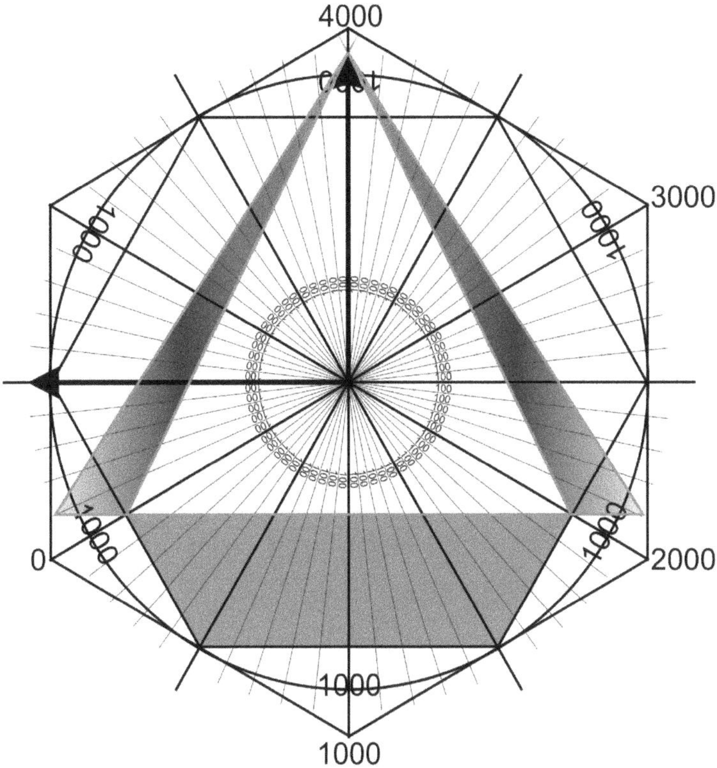

Picture 12 (FLOOD AND LONG GRACE).

This picture confirms the accuracy, that in this period, and in those days of history of this planet, was a flood and occurred in the period around 1.650 Years. Three sons of Noah started a new generation and God had transmitted life through them. Then, this picture shows us that in the period of 2.000 years, on the tenth generation after the flood, from Noah / Nuhh son of Sam; was born Abraham/Ibrahim, through whom God transmitted the bloodline to the birth of Jesus / Issa / Messiah, in the period which is also from 2.000 years.

Then, there was the rainbow of mercy above the Noah's Ark that God created as a sign between people and God. Through it he made a promise that there will be no more floods like this one that flooded the Earth and everything in it and that people should have no more fear of such floods. This picture shows that the rainbow of mercy will exist until the end of life on this planet. The picture shows the first one day or 12 hours, by which we live; then this same clock shows instead of 1 minute -100 years, or 10 minutes -1000 years.

In the middle of the clock, a period or 360 ° circle is shown which confirms the accurate information of the very beginning of time as well as the very beginning of man.

From Adam / Adam to Noah / Nuhh – 1.000 years.

From Nuhh/ Noah to Abraham / Ibrahim – 1.000 years.

From Abraham / Ibrahim to Emperor David / David – 1.000 years.

From Emperor David / David to Jesus / Issa / Messiah – 1.000 years.

Until Christ's Messiah's birth, which was exactly at 4.000 years; it shows time leading us from the first second of counting of time and life on this Earth, at exactly 1.440.000 days life was created on this Earth, the Messiah / Christ was also born.

The picture shows that the (rainbow of mercy) continues in the New Era for as long as there is life on this earth. And where the time that counts in the time of the Old Era ends and where the time of the New Era, in which we live, comes into being.

The picture shows that the (rainbow of mercy) continues in the New Age until

there is life on this Earth.

Mercy continues even after the death of Christ / Messiah death on the cross, until 6.000 years is completed or exactly to the point that connects with the beginning of life on this Earth and where it fits mathematically precisely to zero with the end and at 6.000 years. The day when Adam / Adem sinned God immediately gave his lasting grace for the sinners.

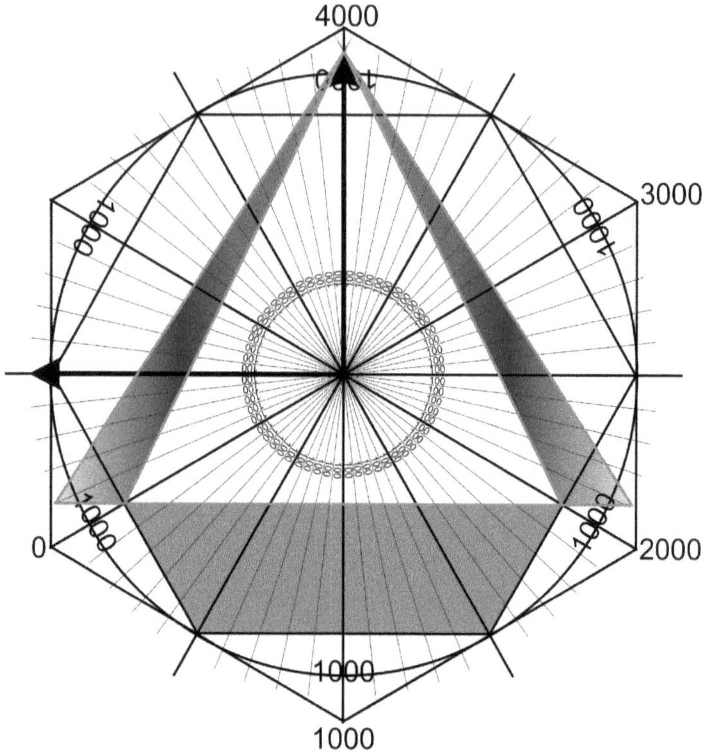

Picture 12a (THE FLOOD IN TIME AND SPACE).

Confirms all this information through time and through the Universe that God revealed his secrets and there is no more secrets other than the secret that will not be revealed and that is Christ / the Messiah's coming.

Because nobody knows the day or hour of Christ's second coming except God. The Angels or Jesus / Issa himself don't know it, that is the secret of God.

Jesus /Issa / Messiah also knows neither the day nor the hour of his coming, which will also be the end of life on this earth, in this sinful form. But everything has its time, and when the time comes at his arrival, he will come just as he came to fulfill the plan o salvation , according to the will of my Father and our Father, who gave everyone by his mercy another chance to get us back to eternal life, where he created us. And who loves us so much, he is patient and through his grace no man shall (perish) to eternal death. We need to seek the path that leads to eternal life.

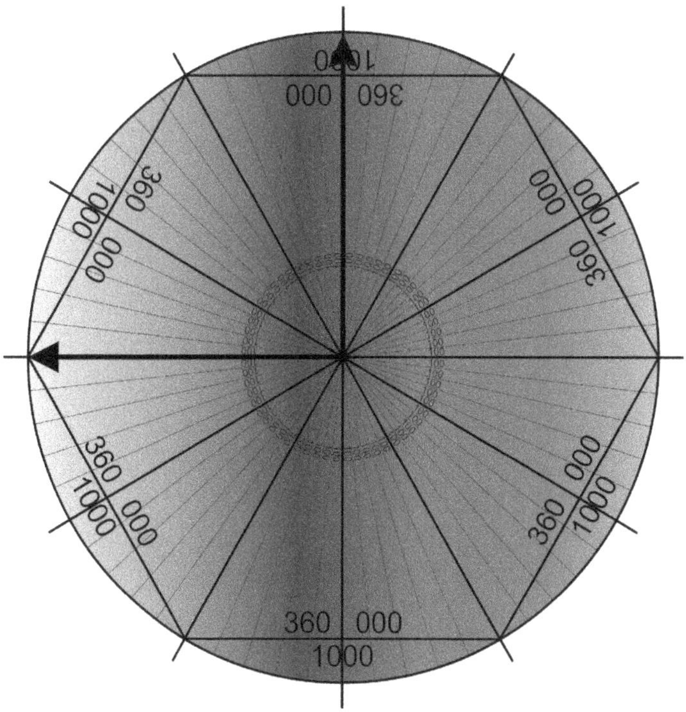

Picture 13 (THE RAINBOW OF MERCY IN ALPHA AND OMEGA).
It also shows a rainbow of mercy that spans the entire planet and which goes through the time of all life on it.

Time through time and the Universe affirms that it is the grace of God that covers all the earth, all people, tribes, and languages. The rainbow of mercy exactly from the cross spans the whole planet and shows that the cross is in the centre of grace, in the centre of space, in the centre of all time, because it is in that place, where it is located the centre of grace, there is also the centre of life, everything created and everything stands on the cross of grace; on Jesus / Messiah. Only he was on that cross and died on it. He confirmed by his day and time of birth that he was the one for whom the entire Universe testifies and for which all numbers and all the letters confirm, which all our genes confirm and show so there would be no doubts and questions as to who is the creator, redeemer and saviour, who is God or if there is one. All the answers are around us, on everything created has the seal of God, but not all can see it because their eyes only see what they want to see. Everyone sees what their heart wants to see and where their desires are.

From these verses we can see that the world because of great sins was flooded and no one survived except Noah and his wife and his three sons and their wives, therefore only eight people survived. And of them God multiplied this world again.

Noah had 600 years, when the flood occurred. Noah had 216.000 days before the flood. Noah lived for another 350 years after the foold and that's another 126.000 days. Altogether, Noah lived 950 years. The flood occurred 1.650 years after Adam / Adem sinned, and had left the Garden of Eden.

In the flood, God washed away the sins committed during that period and shortened life expectancy of people. People lived up to 1.000 years and now peoples life expectancy is about 100 years because of God's grace toward people; for God says: "All the thoughts of our heart are only sin." This means that the longer we live the more sin we make. And for every unforgivable sin, we also receive punishment for sin. The penalty for sin is eternal death.

These pictures show the period of the flood, where it disappeared one whole generation of long-lived life, healthy people which of very good life's brought themselves destruction and eternal death. They didn't feel grateful for the one who gave them all, but perished in their ungrateful behaviour towards the Creator.

Noah / Nuhh hammered out the boat for 120 years and with every sound of the hammer the message was clear that they should stop with their lewd behaviour, to repent and begin to live according to God's instructions and laws.

Picture 14 (BLOOD LINE AND FIRST BORNS) .

It shows, a period of 2.000 years.

The period from the origin of sin on Earth to the Birth of Ibrahim / Abraham.

From Adam / Adem to Noah / Nuha – 1.000 years and 10 generations.

From Noah / Nuha to Abraham / Ibrahim – 1.000 years and 10 generations.

Abraham / Ibrahim is the tenth generation that was specially transmitted, through God's chosen Noah' / Nuh sons.

This information tells us that with God everything is perfect exactly, per second and a minute and fits into everything that God created which is (perfect). This is our instruction that we need to believe that God created everything perfect, even if people are because of their sin, thrown out of the Garden Of Eden. God out of his love for man created a plan of salvation that no one who believes in him will never die but will have eternal life.

From this information, as well as the names of all first-born's heirs there were ten generations from Adam / Adem to Noah / Nuhh.

And from Noah to Abraham / Ibrahim there were also ten generations.

These are the names of the first born who were born after the flood.

Shem, received a blessing from his father and he continued to pass it on to his

bloodline all the way to Abraham / Ibrahim.

These are the generation of Shem:

Shem was 100 years old and begat Ar-phax-ad two years after the flood.

And Shem lived after he begat Ar-phax-ad 500 years, and begat sons and daughters.
Ar-phax-ad lived 35 years and begat Salah:
Ar-phax-ad lived 403 years, begat sons and daughters.
And Salah lived 30 years, begat Eber.
And Salah lived after he begat.
Eber lived 403 years, and begat sons and daughters.
And Eber lived 34 years, and begat Peleg:
And Eber lived after he begat Peleg 430 years, and begat sons and daughters.
And Peleg lived 30 years, and begat Re'-u; and Peleg lived after he begat Re'-u 209 years, and begat sons and daughters.
Re'-u lived 32 years, and begat Serug: And Re'-u lived after he begat Serug 207 years, and begat sons and daughters.
And Serug, lived 30 years, and begat Nahor: And Serug lived after he begat Nahor 200 years, and begat sons and daughters.
And Nahor lived 29 years, and begat Terah: And Nahor lived after begat Terah 190 years, and begot sons and daughters.
And Terah lived 70 years, and begot Abraham, Nahor, in Haren.
From these verses we see that from (Terah) Father of Abraham / Ibrahim were exactly ten generations and 2.000 years, from the beginning of life on the earth.
In the Book of Genesis:
Chapter 12; verse 1. It says,
" And the Lord had said unto Abram, Get thee out of thy country, and from thy kindred, and from thy father's house, unto a land that I will shew thee:

And I'll make of you great people, and I will bless you, and I will glorify your name, and you will be my blessing."

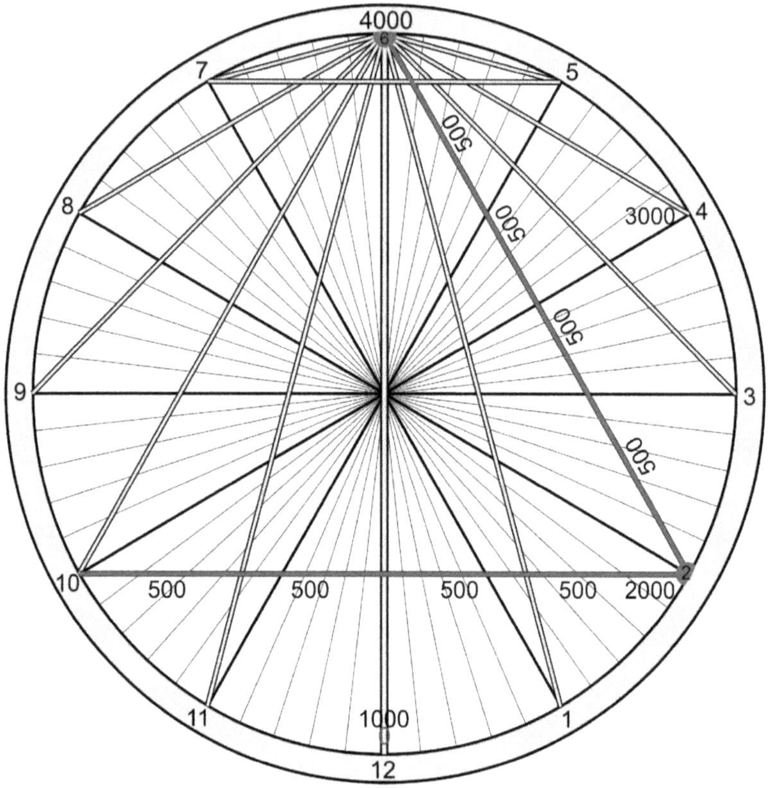

Picture 15 (TIME SPACE AND LIGHT ON THE SAME LINE).

It shows that from Adam / Adam to Abraham / Ibrahim passed exactly 2.000 years, and that is exactly how the Creator separated water from water on the second day of creation. So, on the 2.000 years, he washed the Earth with water, from the sins of Sodom, and new renewed his covenant and bloodline through Abram / Ibrahim and his descendants.

And time shows that, from Adam / Adem to the Flood or Noah / Nuha and to Abram / Ibrahim passed exactly 2.000 years.

Then from Abram / Ibrahim to Emperor David / David was 1.000 years.

From Emperor David / David to Christ / Messiah was 1.000 years. It is the period of 4.000 years through which the bloodline was transmitted.

Abram / Ibrahim, as the Lord told him listened to God and went to the way that God showed him and with him went his brother's son Lot, and Abraham/ Ibrahim was 75 years old when he departed from Haran.

GENESIS: Chapter: 12

Verse 5. "And Abram / Ibrahim took Sarai his wife and Lot his brother's son, and all their substance that they had gathered, and the souls that they had gotten in Haran; and they went forth to go into the land of Canaan; and into the land of Canaan they came.

Chapter 16;

Verse 1. "Sarai Abram's wife bare him no children: and she had an handmaid, an Egyptian, whose name was Hagar."

Verse 2. "And Sarai said unto Abram/ Ibrahim, Behold now, the LORD hath restrained me from bearing: I pray thee, go in unto my maid; it may be that I may obtain children by her. And Abram hearkened to the voice of Sarai."

Verse 4. "And he went into Hagar, and she conceived: and when she saw that she had conceived, her mistress was despised in her eyes."

Verse 5. "And Sarai said unto Abram, My wrong be upon thee: I have given my maid into thy bosom; and when she saw that she had conceived, I was despised in her eyes: the LORD judge between me and thee."

Verse 6. " But Abram said unto Sar-a-I Behold, thy maid is in thy hand; do to her as it pleased thee. And when Sarai dealt hardly with her, she fled from her face."

Verse 7. "And the angel of the LORD found her by a fountain of water in the wilderness, by the fountain in the way to Shur."

Verse 8. "And he said, Hagar, Sarai's maid, whence camest thou? and whither wilt thou go? And she said, I flee from the face of my mistress Sarai."

Verse 9. "And the angel of the Lord said onto her, Return to thy mistress, and submit thyself under her hands."

Verse 10. "And the angel of the Lord said onto her,' I will multiply thy seed exceedingly, that it shall not be numbered for multitude."

Verse 11. "The angel of the Lord still said unto her, Behold, thou art with child, and shalt bear a son, and shalt call his name Ishmael: because the LORD hath heard thy affliction."

Verse 13. "And she called the name of the LORD that spake onto her, Thou God seest me: for she said, have I also here looked after him that seeth me?"

Verse 14. " Wherefore the well was called Beer-la-hai-roi; behold, it is between Kadesh and Bered."

Verse 15. "And Hagar bare Abram a son: and Abram called his son's name, which Hagar bare, Ishmael."

Verse 16. "And Abraham / Ibrahim was 86 years old when Hagar bare Ishmael to Abram."

CHAPTER 17:

Verse 1. "And when Abraham / Ibrahim had 99 years, the LORD appeared to Abram, and said unto him, I am the Almighty God; walk before me, and be thou perfect."

Verse 3. "And Abraham / Ibrahim fell : and God talked with him saying,"

Verse 4. "As for me, behold, my covenant is with thee, and thou shalt be a father of many nations."

Verse 9. "And God said to Abraham / Ibrahim, thou shalt keep my covenant, therefore, thou, and thy seed after thee in their generations."

Verse 10 " This is my covenant, which ye shall keep, between me and you and thy seed after thee; Every man child among you shall be circumcised."

Verse 11. "And ye shall circumcise the flesh of your foreskin; and it shall be a token of the covenant between me and you."

Verse 12. "And he that is eight days old shall be circumcised among you, every man child in your generations, he that is born in the house, or brought with money of any stranger, which is not of thy seed."

Verse 13. "He that is born in thy house, and he that is brought with thy money, must be circumcised: and my covenant shall be in your flesh for an everlasting covenant."

Verse 15. "And God said to Abraham / Ibrahim, as for Sarai thy wife, thou shalt not call shalt her name Sarai, but Sarah shall her name be."

Verse 16. "And I will bless her, and give thee a son also of her: yea, I will bless her, and she shall be a mother of nations; kings of people shall be of her."

Verse 17. "Then Abraham / Ibrahim fell, and laughed, and said in his heart, Shall a child be born unto him that is 100 years old? And shall Sarah, that is 90 years old bear?."

Verse 18. "And Abraham / Ibrahim said unto God, O that Ishmael might live before thee!."

Verse 19. "And God said, Sarah, your wife, will bear thee a son, and you shall call his name is Isaac: and I will establish my covenant with him for an everlasting covenant, and with his seed after him."

Verse 20. "And as for Ishmael I have heard thee: Behold, I have blessed him and will make him fruitful, and will multiply him exceedingly; 12 princess shall he beget, and I will make him a great nation."

Verse 21. "But my covenant will I established with Isaac, which Sarah shall bear unto thee at this set time in the next year."

CHAPTER 21;

Verse 1. "And the LORD visited Sarah as he had said, and the LORD did unto Sarah as he had spoken."

Verse 2. "For Sarah conceived, and bare Abraham/Ibrahim a son in his old age, at the set time of which God had spoken to him."

Verse 3. "And Abraham / Ibrahim called the name of his son that was born unto him, whom Sarah bare to him, Isaac."

CHAPTER 23.

Verse 1. "And Sarah lived 127 years; these were the years of the life of Sarah."

Verse 2. "And Sarah died in the city of Kirjath-arba; and that is Hebron in the land Ca-na-an."

CHAPTER 24.

Verse 1. "Abraham / Ibrahim was old, and well stricken in age: and the LORD had blessed Abraham/Ibrahim in all things."

Verse 2. "And Abraham / Ibrahim said unto his eldest servant of his house, that ruled over all that he had, Put, I pray thee, thy hand under my thigh."

Verse 3. "And I will make thee swear by the LORD, the God of heaven, and the God of the earth, that thou shalt not take a wife unto my son of the daughters of the Canaanites, among whom I dwell."

Verse 4. " But thou shalt go into my country, and to my kindred, and take a wife unto my son, Isaac."

Verse 9. "And the servant put his hand under the thigh of Abraham/Ibrahim his master, and sware to him concerning that matter."

Verse 10. "Then the servant took ten camels of the camels of his master, and departed; for all the goods of his master were in his hand: and he arose, and went to Mesopotamia, unto the city of Nahor."

Verse 12. "He said, O LORD God of my master Abraham / Ibrahim, I pray thee, send me good speed this day, and shew kindness unto my master, Abraham / Ibrahim."

Verse 13. "Behold, I will stand here by the well of the water; and the daughters of the men of the city come out to draw water."

Verse 14. "And let it come to pass, that the damsel to whom I shall say, let down thy pitcher, I pray thee, that I may drink; and she shall say, Drink, and I will give thy camels drink also; let the same she that thou hast appointed for thy servant Isaac; and thereby shall I know that thou hast shewed kindness unto my master."

Verse 15. "And it came to pass, before he had done speaking, that ,behold, Rebekah came out, who was born to Bethuel, son of Milcah, the wife of Nahor, Abrahams brother, with her pitcher upon her shoulder."

Verse 16. "And the damsel was very fair to look upon, a virgin, neither had any man known her: and she went down to the well, and filled her pitcher, and came up."

Verse 17. "And the servant ran to meet her, and said, Let me, I pray thee, drink a little water of thy pitcher."

Verse 18. "And she said, Drink, my lord: and she hasted, and let down her pitcher upon her hand, and gave him drink."

Verse 19. "And when she had done giving him drink, she said, I will also draw water for thy camels also, until they have done drinking."

Verse 22. "And when the camels had done drinking, that the man took a golden earring of half a shekel weight and two bracelets for her hands of ten shekels weight of gold."

Verse 23. "And he said, 'Whose daughter art thou? Tell me, I pray thee: is there room in thy father's house for us to lodge in."

Verse 24. "And she said unto him, I am the daughter of Bethuel the son of Milcah which she bare unto Nahor."

Verse 26. "And the man bowed down his head, and worshipped the LORD."

Verse 34. "Then he said, I am the servant of Abraham."
Verse 49. "And now if ye will deal kindly and truly with my master, tell me: and if not, tell me; that I may turn to the right hand, or to the left."
Verse 51. "Behold, Rebekah is in your power, take her, and go, and let her be a wife to your master's son, as the Lord says."
Verse 52. "And it came to pass, when Abraham's / Ibrahim servant heard their words, he worshipped the LORD, bowing himself to the Earth."
Verse 58. "And they called Rebekah, and said unto her, Will you go with this man? And she said, I will go."
Verse 63. "And Isaac went out into the field the night to pray to God. Raising his eyes he saw the camels were coming."
Verse 66. "And Isaac's servant told him all that he had done."
Verse 67. "And Isaac brought her into his mother Sarah's tent, and took Rebekah, and she became his wife; he loved her: and Isaac was comforted after his mother's death."
CHAPTER 25;
Verse 7. "And these are the days of the years of Abraham/Ibrahim life which he lived 175 years."
Verse 8. "And Abraham / Ibrahim died in good old age, an old man, and full of years; and was gathered to his people."
Verse 9. "And Isaac and Ishmael buried him in the cave of Machpelah."
Verse 11. "And after the death of Abraham / Ibrahim, God blessed his son Isaac; And Isaac dwelt by the well La'-hai-roi."

Verse 12. “Now these are the generations of Ishmael, Abraham’s / Ibrahim son, whom Hagar, the Egyptian, Sarah’s handmaid, bare unto Abraham /Ibrahim.”

Verse 13. “And these are the names of the sons of Ishmael, by their names, according to their generations. The firstborn of Ishmael: Ne-ba-joth; and Kedar, and Adbeel, and Mibsam.”

Verse 14. “And Mishma , and Dumah, and Massa.”

Verse 15. “And Hadar, And Tema, And Jetur, And Nafis, and Kedmah.”

Verse 16. “These are the sons of Ishmael; these are their names in the villages, and their castles. Twelve princes according to their nations.”

Verse 17. “And these are the years of the life of Ishmael, an 137 years and he gave up the ghost and died.”

Verse 18. “And they dwelt from Havilah unto Shur, that is before Egypt, as thou goest toward Assyria.”

And these are the generations of Isacc the son of Abraham / Ibrahim. Abraham / Ibrahim begat Isaac.

Verse 20. “And Isaac was 40 years old when he married Rebekah, the daughter Bautehuel the Syrian of Padam-ara, the sister to Laban the Syrian.”

Verse 21. “And Isaac prayed unto the Lord for his wife, because he was barren and the LORD was entreated of hum, and Rebekah his wife conceived.”

Verse 22. “And the children struggled together within her; and she said, if it be so, why am I thus? And she went to inquire of the LORD.”

Verse 23. "And the Lord said unto her, There are two nations in thy womb, and two manner of people shall be separated from thy bowels. One people will be stronger than the other people and the elder shall serve the younger."

Verse 24. "And when her days to be delivered were fulfilled, there were twins in her womb."

Verse 25. "And the first went out red, all over like an hairy garment, and called his name Esau, and they named the other Jacob / Jakub. And Isaac was 60 the year when Rebecca gave birth to them."

Verse 28. "And Isaac loved Esau, because he did eat of his venison: but Rebekah loved Jacob / Jacob."

Verse 29. "And Jacob sod pottage and Esau came out of the field faint."

Verse 30. "And Esau said to Jacob / Jacub, Feed me I pray thee, with that same red pottage; for I am faint: therefore was his name called Edom."

Verse 31. "And Jacob / Jacub said to him, Sell me this day thy birthright."

Verse 32. "And Esau said, I am at the point to die; and what profit shall this birthright do to me?."

Verse 33. "Jacob said to him, Swear to me this day; and he sware unto him: and he sold his birthright unto Jacob / Jakub."

Verse 34. "And Jacob / Jacub gave Esau bread and pottage of lentils, and he ate and drank and then got up and left. Esau despised his birthright."

CHAPTER 27:

Verse 1. "And when Isaac was old and his eyes were dim, so that he could not see, he called Esau his eldest son, and said unto him, Behold, here I am."

Verse 2. "Then he said, Behold now, I am old; I know not the day of my death."

Verse 3. "Now therefore take, I pray thee, thy weapons, thy quiver and thy bow, and go out to the field, and take me some venison."

Verse 4. "And make me savoury meat, such as I love, and bring it to me, that I may eat, that my soul may bless thee before I die."

Verse 5. "And Rebecca heard when Isaac spake to Esau his son. And Esau went to the field to hunt for venison, and to bring it."

Verse 6. "Rebekah spake unto Jacob / Jacub her son, Behold, I heard thy father speak unto Esau thy brother, saying."

Verse 7. "Bring me venison, and make me savoury meat, that I may eat, and bless thee before the LORD before my death."

Verse 8. "Now, therefore my, son, obey my voice according to that which I command thee."

Verse 9. "Go now to the flock , and bring two good goats ; and I will make them savoury meat for thy father, such as he loveth."

Verse 10. "And bring it to thy father, that he may eat, and that he may bless thee before his death."

Verse 11. "And Jacob / Jacub said to Rebekah his mother, But Esau, my brother is a hairy man, and I am a smooth man."

Verse 12. "My father peradventure will feel me, and I shall seem to him as a deceiver; and I shall bring a curse upon me, and not a blessing."

Verse 13. "And his mother said unto him, Upon me be thy curse, my son: only obey my voice, and go fetch me them."

Verse 14. "Then he went and fetched, and brought them to his mother, and his mother made savoury meat, such as his father loved."

Verse 15. "And Rebekah took goodly raiment of her eldest son Esau, which were with her in the house, and put upon Jacob / Jakub her younger son."

Verse 16. "And she put the skins of the kids of the goats upon his hands, and upon the smooth of his neck."

Verse 17. "And she gave the savoury meat and the bread, which she had prepared, into the hand of her son Jacob / Jakub."

Verse 18. "And he came unto his father, and said, my father: and he said, here I am, who art thou, my son?."

Verse 19. "And Jacob / Jacub said unto his father, I' am Esau, thy firstborn: I have done according as thou badest me; arise, I pray thee, sit and eat of my venison, that thy soul may blees me."

Verse 20. "And Isaac said unto his son, How is it that thou hast found it so quickly, my son? And he said, Because the LORD thy God brought it to me."

Verse 21. "Then Isaac said unto Jacob / Jacub, Come closer, son, that you I feel whether you are my son Esau or not."

Verse 22. "And Jacob / Jacub came to his father, and he felt him and said, 'The voice is Jacob's / Jacub's, but the hands are the hands of Esau."

Verse 23. "And he discerned him not, because his hands were hairy, as his brother Esau's, hands: So he blessed him."

Verse 26. "And his father Isacc said unto him, Come, near now, and kiss me, my son."

Verse 30. "And it came to pass, as soon as Isaac had made an end of blessing Jacob / Jacub and Jacob / was yet scarce gone out from the presence of Isaac his father, that Esau his brother came in from his hunting."

Verse 31. "And he also made savoury meat, and brought it unto his father, and said unto his father, let my father arise, Arise, and eat of his son's venison, that thy soul may bless me."

Verse 32. "And Isaac his father said unto him, Who art thou? And he said, I 'm thy son, the firstborn Esau."

Verse 33. "And Isaac trembled very exceedingly, and said, Who? Where is he that hath taken venison, and brought it me, and I have eaten of all before thou camest, and have blessed him ? yea and he shall be blessed."

Verse 34. "And when Esau shall heard the words of his father, he cried with a great and exceeding bitter cry, and said unto his father, Bless me, even me also."

Verse 35. "And he said, Thy brother came in with subtility, and hath taken away thy blessing."

Verse 41. "And Esau hated Jacob / Jacub because of the blessing where with his father blessed him, and Esau said in his heart, The days of mourning for my father are at hand; then will I slay my brother Jacob / Jacub."

Verse 42. "And these word of Esau her elder were told to Rebekah: and she sent and called Jacob, her younger son and said unto him, Behold, thy brother Esau, as touching thee, doth comfort himself, purposing to kill thee."

Verse 43. "Now therefore, my son, obey my voice, and arise, flee thou to Laban my brother to Haran."

CHAPTER 28

Verse 7. "And Jacob / Jacub obeyed his father and his mother, and was gone to Padam -Aram."

Verse 10. "Jacob / Jacub went out from Beer-sheba, and went toward Haran."

Verse 11. "And he lightened upon a certain place, and tarried there all night, because the sun was set; and he took of the stones that place, and put them for his pillows, and lay down in that place to sleep."

Verse 12. "And he dreamed, and behold a ladder set up on the earth, and the top of it reached to heaven and behold the angels of God ascending and descending on it."

Verse 18. "And Jacob / Jacub rose up early in the morning, and took the stone that he had put for his pillows, and set it up for a pillar, and poured oil upon the top of it."

Verse 19. "And he called that the name of that place Bethel: but the name of that city was called Luz at the first."

CHAPTER 29;

Verse 1. "Then Jacob / Jacub went on his journey, and came into the land of the people of the east."

Verse 2. "And he looked, and behold a well in the field and , lo, there were three flocks of sheep are lying by it; for out that well they watered the flocks: and a great stone was upon the well's mouth."

Verse 3. "And thither were all the flocks gathered: and they rolled the stone from the well's mouth, and watered the sheep, and put the stone again upon the well's mouth in his place."

Verse 4. "And Jacob / Jacub said unto them, My brethren, whence be ye? And they said, Of Haran are we."

Verse 5. "And he said unto them, Know ye Laban the son of Nahor?" And they said, 'We know him."

Verse 6." And he said to them, Is he well? And they said, he is well: and behold, Rachel his daughter cometh with the sheep."

Verse 11. "And Jacob/ Jakub kissed Rachel, and lifted up his voice, and wept."

"And Jacob / Jacub married in Haran, and took Rachel and Leah for women. And his wives slaves, which their father gave them when they married. They birth to Jacob / Jacub eleven sons and one daughter in Padam-aram and the Lord spoke to Jacob / Jacub.

Verse 13. "I am the God of Bethel, where thou hast anointedst the pillar, and where thou vowest a vov unto me: now arise, get thee out from this land country and return unto the land of thy kindred."

"Then Jacob / Jakub, his wives, and his children, and all his servants, and all his cattle, and all that was gained in Padam -aram for the twenty years he worked with his father-in-law for the salary. They moved back to their stable. For the Lord said to Jacob / Jacub, return to the land of your fathers and your kind and I will be with you."

CHAPTER 32.

Verse 1. "And Jacob / Jacub went his way, and the angels of God met him."

Verse 3. "Jacob / Jacub sent messengers before him to Esau, his brother unto the land Seir, the country of Edom."

Verse 4. "And he commanded them, saying, Thus shall ye speak unto my lord Esau: Thy servant, Jacob / Jacub saith thus, I have sojourned with Laban, and stayed there until now:"

Verse 5. "And I have oxen and asses, flocks, men servants, and women servants, and I have sent to tell my lord, that I may find grace in thy sight."

Verse 6. "And the messengers returned to Jacob / Jacub, saying. We came to thy brother Esau, and also he cometh to meet thee, and four hundred men with him."

Verse 7. "And Jacob / Jacub was greatly afraid and distressed: and he divided the people that was with him, and the flocks, and the herds, and the camels, into two bands."

Verse 8. "And said, If Esau come to the one company, and smite it, then the other company which is left shall escape."

Verse 9. "And Jacob said, O God of my father Abraham /Ibrahim, and God of my father Isaac, the LORD which sadist unto me, Return unto thy country, and to thy kindred, and I will deal well with thee."

Verse 10. "I am not worthy of the least of all the mercies, and of all the truth, which thou hast shewed unto thy servant; for with my staff I passed over this Jordan; and now I am become two bands."

Verse 11. "Deliver me, from the hand of my brother, from the hand of Esau: for I fear him, lest he will come and smite me , and the mother of my children. "

Verse 13. "And he lodged there that same night; and took of that which care to his hand a present for Esau his brother."

Verse 14. "Two hundred she goats with twenty he goats, two hundred sheep and twenty rams."

Verse 15. "Thirty milch camels with their colts, forty kine, and 10 bulls, twenty she asses, and 10 foals."

Verse 21. "So went the present over before him: and himself lodged that night in the company."

Verse 22. "And he rose up that night, he took his two wives, and his slaves and his eleven sons , and passed over the ford Jabbok."

Verse 24. "And when Jacob / Jacub was alone, one man wrestled with him till dawn."

Verse 25. "And when he saw that he prevailed not against him, he touched the hollow of his thigh; and the hollow of Jacob's thigh was out of joint, as he wrestled with him."

Verse 26. "Then he said, Let me go, it's dawn. And Jacob / Jacub said, I will not let you go until you bless me."

Verse 27. "And he said unto him, What is thy name?" And he answered, Jacob / Jacub. "And he said. Thy name shall be called no more Jacob/ Jacub, but ISRAEL: for as a prince hast thou power with God and with men, and hast prevailed."

Verse 29. "And Jacob / Jacub asked him, and said, Tell me, what's your name? And he said, 'why you ask my name? And blessed him there."

CHAPTER 33:

Verse 1. "And Jacob / Jacub lifted up his eyes, and Esau came, and with him four hundred men. And he divided the children unto Leah, and unto Rachel, and unto the two handmaids."

Verse 4. "And Esau ran to meet him, and embraced him, and fell on his neck, and kissed him: and they wept."

Verse 16. "So Esau returned that day on his way unto Seir."

Verse 18. "After Jacob / Jacob came to Shalem, a city of She-chem, which is the land of Canaan, when he came from Padanaram; and pitched his tent before the city."

Verse 20. "And he erected there an altar, and called it Elelohe Israel."

CHAPTER 35:

Verse 1. "And God said unto Jacob / Jacub, Arise, go up to Bethel, and dwell there: and make there an altar unto God, that appeared unto thee when thou fleddest from the face of Esau thy brother."

Verse 6. "And Jacob / Jacub, came to Luz, which is in the land of Canan, that is, Bethel, he and all the people that were with him."

Verse 7. "And he built there an altar, and called it the place El-beth-el: because there God appeared unto him, when he fled from the face of his brother."

Verse 9. "And God appeared unto Jacob / Jacub again, when he came out of Padan-aram and blessed him."

Verse 10. "And God said unto him, Thy name is Jacob / Jacub: thy name shall not be called any more Jacob/ Jacub, but ISRAEL shall be thy name; and he called his name ISRAEL."

Verse 12. “And the land which I gave Abraham / Ibrahim and Isaac, to thee I will give it, and to thy seed after thee will I give the land.”
Verse 15. “And Jacob / Jacob called the name of the place where God spake with him, Beth-el.”
Verse 16. “And they journeyed from Beth-el; and there was but a little way to come to Ephrath: and Rachel travailed, and she had hard labour.”
Verse 17. “And it came to pass, when she was in hard labour, that the midwife said unto her, Fear not; thou shalt have this son also.”
Verse 18. “And it came to pass, as her soul was in departing, for she died that she called his name Ben-o-ni: but his father called Benjamin.”
Verse 19. “And Rachel died, and was buried in the way to Ephrath, which is Beth-lehem.”
Verse 20. “And Jacob / Jacub set a pillar upon her grave: that is the pillar of Rachel’s grave unto this day.”
Verse 27. “Jacob / Jacub also came unto Isaac his father unto Mamre, unto the city of Arbah, which is Hebron, where Abraham / Ibrahim and Isaac sojourned.”
Verse 28. “And Isaac was 180 years old, and he died. And his sons Esau and Jacob buried him.”
CHAPTER 37:
Verse 1. “And Jacob / Jacub lived in the land where his father was a stranger. In the land of Canaan.”
Verse 3. “And Israel loved Joseph / Yusuf more than all his children, because he was the son of his old age: and he made him a coat of many colours.”

2. The Book of Exodus

CHAPTER 1:

These are the names of the children of Israel who came to Egypt. They came with Jacob / Jacub, each with his own family: Reuben, Simon, Levi, Judah, Issachar, Zabulon, Benjamin, Dan, Naphtali, Gad, and Asar.

Twelve sons of Jacob / Jacub who came into Egypt with their families, and there were seventy souls.

All of them were from the loins of Jacob, seventy souls with Joseph / Yusuf. Which was in Egypt. And Joseph / Yusuf died, and all his brothers and all that generation. And the children of Israel were born also they multiplied, prospered, and strengthened themselves very much and filled up the land.

Verse 8. “Now there arose up a new king over Egypt, which knew not Joseph/ Yusuf.”

Verse 9. “And he said unto his people, Behold, the people of the children of Israel, are more and mightier than we.”

Verse 10. “But let us deal wisely with them, lest they multiply, and it come to pass, that, when there falleth out any war, they join also unto our enemies, and fight against us, and so get them up out of the land.”

Verse 15. “And the king of Egypt commanded the midwives of the Hebrews, and said ‘When you babysit a Jew and if it be a son, kill him: but if it be a daughter , then she shall live.”

Verse 17. “But the midwives feared God, and did not as the king of Egypt command them, but saved the men children alive.”

Verse 22. “Then Pharaoh commanded his people, saying, Every son that is born ye shall cast into the river, and every daughter ye shall save alive.”

CHAPTER 2:

Verse 1. “And one of the tribes of Levi went and married the daughter of Levi.”

Verse 2. “And she became pregnant, giving birth to a son, and seeing he was a godly child, she hid him three months.”

Verse 3. “And when she could no longer hide him, she took for him an ark of bulrushes, and daubed it with slime and with pitch, and put the child therein; and she said it in the flags by the river’s brink.”

Verse 4. “And his sister stood afar off, to see what would be done to him.”

Verse 5. “And Pharaoh’s daughter came to bathe in the river, and her maidens they were walking by the river, and when she saw the ark among the flags, she sent her maid to fetch it.”

Verse 6. “And when she opened it , she saw the child, and behold, the child wept. And she had compassion on him, and said, this is a Jewish child.”

Verse 7. “Then said his sister to Pharaoh’s daughter, Shall I go and call to thee a nurse of the Hebrew women, that she may nurse the child for thee?.”

Verse 8. “And Pharaon’s daughter said to her, Go. And the maid went and called the child’s mother.”

Verse 9. “And Pharaon’s daughter said unto her, Take this child away and nurse it for me, and I will give thee thy wages. And the woman took the child , and nursed it.

Verse 10."And when the child grew, he was taken to Pharaoh's daughter, and she gave him the name Moses / Musa."

Verse 11. "When Moses / Musa was grown, he went out unto his brethren and looked on their burdens. And saw where an Egyptian was beating a Jew man among his brethren.

Verse 12."And he looked around. When he saw that there was no one, he killed the Egyptian and buried him in the sand.''

Verse13"The next day he went out again, and the two Jews argued, and said to him who were doing wrong thing, Why are you beating your neighbour?'

Verse 14"And he said, Who made you a prince and judge over us? Do you want to kill me are you as you killed the Egyptian? 'Then Moses / Musa got frightened and he said, It has been revealed!'

Verse 15."And Pharaoh, hearing this, looked for Moses / Musu. But Moses / Musa fled from Pharaoh and went to the land of Midian, and they sat at one well."

Verse 16."And the priest of Midian had seven daughters, and they came to draw water.

The shepherds came and they were driven away, but Moses / Musa stood up and helped them. And when they returned to Reu'el their father and said: An Egyptian delivered us out of the hand of the shepherds, and also drew water enough for us, and watered the flock.

And he said to his daughters, Where is he? Why have you left this man? Call him to eat!."

Verse 21. "Moses / Musa tended was content to dwell with the man: and gave his daughter Zip-po-rah to him. She gave him a son, and he gave him his name Gershom. After a long time the Egyptian Emperor died, they sighed out of distress, the children of Israel cried out. They cried, their cry came up unto God by reason of the bondage. And God heard their groaning, and God remembered his covenant with Abraham, Isaac and with Jacob / Jacub, and God, looked upon the children of Israel , and had respect unto them."

CHAPTER 3:

Verse 1. "Now Moses / Musa kept the flock of Jethro his father-in-law, the priest of Midian: and he led the flock to the backside of the desert, and came to the mountain of God, even to Horeb."

Verse 2. "And the angel of the LORD appeared unto him in a flame of fire out of the midst of a bush: and he looked, and, behold, the bush burned with fire, the bush was not consumed."

Verse 3. "And Moses/Musa said, I will now turn aside, and see this great sight, why the bush is not burnt."

Verse 4. "And when the LORD saw that he turned aside to see, God called unto him of the midst of the bush, and said, Moses/ Musa . And he said, Here am I."

Verse 5. "God said, Don't come here. Take off your shoes from your feet, for where you stand is a holy land."

He said, I am the God of your father, the God of Abraham / Ibrahim, the God of Isaac, and the God of Jacob. And Moses hid his face; for he was afraid to look upon God.

Verse 10. "Come now therefore, and I will send thee unto Phar'aoh, that thou mayst bring forth my people the children of Israel out of Egypt. And Moses said upon to God, Who am I? I, that should go unto Phar,aoh, and that I should bring forth the children of Israel out of Egypt?

Verse 12. "God said to him, I' will be with you and let this be your token that I sent you. When you bring the people out of Egypt, you will serve God on this mountain."

Verse 15. "And again he said unto Moses, Thus say ye unto the children of Israel, The Lord God, of Abraham / Ibrahim, the God of Isaac, and the of Jacob, hath sent me unto you: this is my name for ever, and this is my memorial unto all generations."

CHAPTER 4:

Verse 1. "Moses / Musa answered and said, But they will not believe me, neither listen to my voice because they will say, the Lord hath not appeared unto thee."

Verse 11. "The Lord said to him, Who made man's mouth? Who can create dumb or deaf or blind? Isn't it me? Have not I the LORD? Go, I will be with your mouth and I will teach you what you will say."

Verse 27. "The Lord said to Aaron / Harun, Go out into the wilderness to meet Moses / Musa. And they went and met him in the mount of God, and Kissed him."

Verse 28. " And Moses/Musa told Aaron / Harun his brother all the words of the LORD who had sent him and signs which he had commanded him."

Verse 30. "And Aaron / Harun spake all the words which the LORD had spoken unto Moses / Musa, and did the signs in the sight of the people."

Verse 31. "And the people believed: and when they heard that the LORD had visited the children of Israel and that he had looked upon their affliction, then and they bowed their heads, and worshipped."

CHAPTER 5:

Verse 1. "And Moses / Musa and Aaron / Harun came out before Pharaoh, and they said to him, Thus says the LORD, God of Israel, let my people go, that they may hold a feast unto me in the wilderness."

Verse 2. "Pharaoh said, 'Who is the LORD, that I should obey his voice and let Israel go? I do not know the LORD, nor will I let Israel go!."

CHAPTER 6:

Verse 1. "Then the LORD said, to Moses, Now you will see what I will do to Pharaoh: for with a strong hand shall he drive out of his land."

Verse 2. "And God spake unto Moses / Musa, and said unto him, I'm the" LORD."

Verse 3. "And I appeared unto Abraham / Ibrahim, unto Isaac and unto Jacob/Jacub by the name of (God Almighty). "but by my name JEHOVAH was I not known to them."

Verse 5. "And I heard the groaning of the children of Israel which the Egyptians hold in to slavery and I have remembered my covenant."

Verse 8. " And I will bring you in unto the land, concerning the which I did swear to give it to Abraham / Ibrahim, to Isaac and to Jacob / Jacub; and I will give it you for an heritage: I am the LORD."

CHAPTER 11:

Verse 1. "And the LORD said unto Moses / Musa, Yet will I bring one plague more upon Pharaoh, and upon Egypt; afterwards he will let you go hence: when he shall let you go, he shall surely thrust you out hence altogether."

Verse2. "Speak now in the ears of the people, and let every man borrow of his neighbour, and every woman of her neighbour, jewels of silver, and jewels of gold."

Verse 3. "And the LORD gave the people favour in the sight of the Egyptians. Moreover the man Moses/Musa was very great in the land of Egypt, in the sight of Pharaohs servants, and in the sight of the people."

Verse 4. "And Moses said, Thus said the LORD, About midnight will I go out into the midst of Egypt."

Verse 5. "And all the firstborn in the land of Egypt shall die, from the Pharaoh that sitteth upon his throne, even unto the firstborn of the maidservant that is behind the mill; and all the firstborn of beasts."

CHAPTER 12:

Verse 1. "The Lord spoke undo Moses / Musa, and Aaron / Harun, in the land Egypt saying,"

Verse 2. "This month shall be unto you the beginning of months: it shall be the first month of the year to you."

Verse 3. "Speak ye unto all the congregation of Israel, In the tenth day of this month they shall take to them every man a lamb, according to the house of their fathers a lamb, for an house."

Verse 4. “And if the household be too little for the lamb, let him and his neighbours next unto his house take it according to the number of the souls; every mam according to his eating shall make your count for the lamb.”

Verse 5. “Your lamb shall be without blemish, a male of the first year.”Verse 6. “And ye shall keep it up until the fourteenth day of this month: and the whole assembly of the congregation of Israel shall kill it in the evening.”

Verse 7. “And they shall take of the blood, and strike it on the two side posts and on the upper door post of the house, wherein they shall eat it.”

Verse 8. “Let them eat flesh the same night, roasted with fire, with unleavened bread and with bitter herbs. Don’t leave anything till morning, if what would be left until morning, burn with fire.”

Verse 11. “And thus shall ye eat it; with your loins girded, your shoes on your feet, and your stick in your hand; and ye shall eat it in haste: it is the LORD’s Passover.”

Verse 14. “And this day shall be unto you for a memorial; and ye shall keep it a feast to the LORD throughout your generation; ye shall keep it a feast by an ordinance for ever.”

Verse 15. “Seven days shall ye eat unleavened bread; even the first day ye shall put away leavened bread. Whosoever eats leavened bred from the first day, until the seventh day, that soul shall be cut off from Israel.”

Verse 16. “And the first day there shall be a holy convocation, and in the seventh day there shall be an holy convocation to you ; no manner of work shall be done in them, save that which every man must eat, that only may be done of you."

Verse 18. "In the first month of the 14th day start eating unleavened bread, until the one and twentieth day of the month at even." And the sons of Israel dealt in Egypt 430 years. And when 430 came to a close on the same day all the armies of the LORD came forth out of the land of Egypt."

When the time came for the LORD to bring them out of Egypt, as he promised to their fathers.

When he brought Abraham / Ibrahim from Mesopotamia, from Haran and they brought him to the land which he had promised, the land of Canaan, and told him that he would give that land to him and his tribe.

But then he told him that his sons will be strangers and slaves in Egypt for 400 years, so after that they will bring them out and bring them back to the land of Canaan."

CHAPTER 13:

Verse 19. "Moses / Musa took the bones of Joseph / Yusuf with him. For he had sworn to the children of Israel, for God will visit you, and ye shall carry up my bones away hence with you."

And when it was reported to the Emperor of Egypt that the people had fled: and the heart of Phar,-aoh and of his servants was turned against the people, and they said, "Why have we done this, that we have let Israel go from serving us?" And he took six hundred chosen chariots and all the chariots of Egypt, and captains over every one of them."

CHAPTER 1. Verse 10. "
When Pharaoh drew nigh, the children of Israel lifted up their eyes, and the Egyptians marched after them; and the children of Israel cried out unto the LORD."
Verse 13. " And Moses / Musa said unto the people, Fear ye not, stand still and see the salvation of the LORD which he will shew to you to – day: for the Egyptians whom ye have seen to-day, ye shall see them again no more for ever."
Verse 15. "The LORD said unto Moses, Wherefore criest thou unto me? Speak unto the children of Israel, that they go forward."
Verse 16. "But lift thou up thy rod, and stretch out thine hand over the sea, and divide it: and the children of Israel shall go on dry ground through the midst of the sea."

Verse 21. "And Moses / Musa stretched out his hand over the sea; and the LORD caused the sea to go back by a strong east wind all that night, and made the sea dry land, and the waters were divided."
Verse 22. "And children of Israel went into the midst of the see upon the dry ground: and the waters were a wall unto them on their right hand, and on their left."
Verse 23. "And the Egyptians pursued, and in after them to the midst of the sea, even all Pharaoh's horses, his chariots, and his horsemen."
Verse 24. "And it came to pass that in the morning watch the LORD looked unto the host of the of the Egyptians through the pillar of fire and of the cloud, and troubled the host of the Egyptians."

CHAPTER 19:

Verse 1. “On the first day of the third month, after the children of Israel had come out of Egypt, they came to the desert of Sinai.”

Verse 10. “And the LORD said unto Moses / Musa, Go unto the people, and sanctify them, today and tomorrow, and let them wash their clothes.”

Verse11. “And be ready against the third day: for the third day the LORD will come down in the sight of all the people upon mount Sinai.”

Verse 14. “And Moses / Musa went down from the mount unto the people, and sanctified the people; and they washed their clothes.”

Verse 15. “And he said unto the people, be ready against the third day: come not at your wives.”

Verse 25. “So Moses / Musa went down unto the people, and spake unto them.”

The Book of Genesis or Book 2 of Moses:

CHAPTER 20:

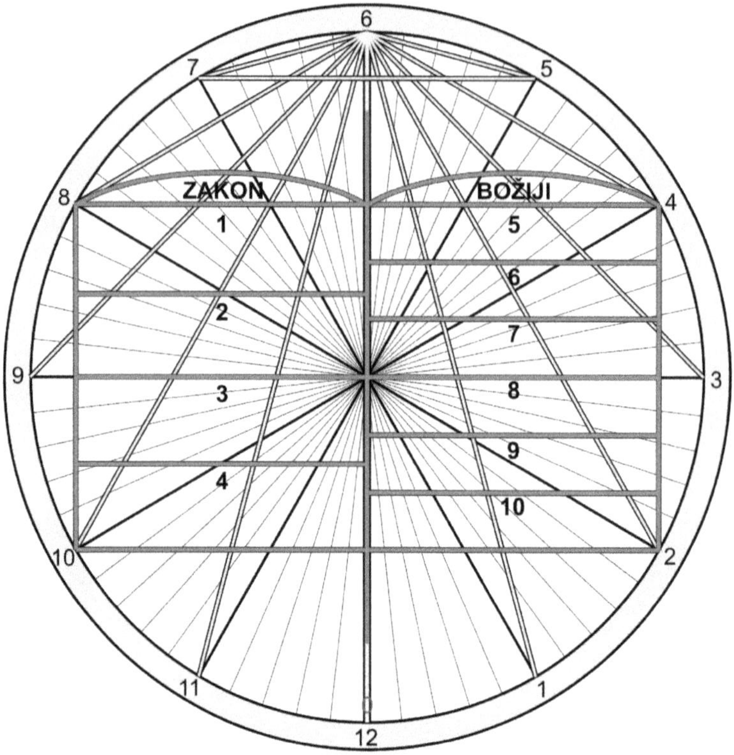

Pciture 16 (THE LAW OF GOD).

The law of God given to Moses / Musa at Mount Sinai.

An hour, an hour of light, an hour of grace, an hour of God revealed to us everything visible and invisible showed people that it was true, in time coming out of Egypt, the LORD: gave His law, the Ten God's commandments: He gave to Moses / Musa on Mount Sinai. Who is the LORD – GOD: wrote with his finger and delivered it into the hands of Moses / Musa, the law eternal, by which men will be held accountable for their works and by whom they are judged for eternal death or eternal life.

This perfect circle, Alpha and Omega, in which, through time and space, the law of God showed to be in the centre through the rays of sun and light illuminate it. They point the great importance of the law of God. The universe, space and time witnesses that we will all be condemned to him.

The law reveals obedience, respect, recognition and faith in God, as the only one to whom all glory belongs, and thanks for all that we have and what we will have. As well as acknowledging that he alone is the only one. He created everything and there is no other God but him. The law of God accurately describes who created heaven and earth, life on earth and for how many days. So no one can have an excuse not to have information or that he did not know.

EXODUS: CHAPTER 20:

Verse 1. " And GOD SPAKE ALL THESE WORDS SAYING: "

1).The first commandment is:

I 'am the LORD thy God, which have brought thee out of the land of Egypt, out of the house of bondage.

2).The second commandment: Though shalt have no other gods before me.

2). The third commandment: Thou shalt not make unto thee any graven image, or any likeness of anything that is in heaven above, or that is in the earth beneath, or that is in the water und earth: Thou shalt not bow down thyself to them, nor serve them: for I the LORD thy God am a JEALOUS God, visiting the iniquity of the fathers upon the children unto the third and fourth generation of them that hate me; And shewing mercy unto thousands of them love me, and keep my commandments. Thou shalt not take the name of the LORD thy God in vain; for the LORD will not hold him guiltless that taketh his name in vain.

4).The fourth commandment: Remember the SABBATH day, TO KEEP IT HOLY. Six days shalt thou labour, and do all thy work: But the seventh day is the Sabbath of the LORD thy God: in it thou shalt not do any work, thou, nor thy son, nor thy daughter, thy manservant, nor thy maidservant, nor thy cattle, nor thy stranger that is within thy gates. For in six days the LORD made heaven and earth, and rested the seventh day: wherefore the LORD blessed the SABBATH day, and hallowed it.

These are the first four commandments of God concerning the LORD - GOD

They describe who created the earth and the sky and all that is seen and not seen. He pointed us in the direction and explained who has the power to create it all. He showed that statues, nor any idols, did not create anything and that only he is the one who created everything visible and invisible around us, he gives and takes. In his hands is life and death. He alone is God, there is no other God but him.

These are the six commandments of God that are on the second plate Of the LOW of God

There are ten in total.

---

5). Fifth Commandment:

“Honour thy father and thy mother: that thy days be long upon the land which the LORD thy God giveth thee.”

6). Sixth Commandment:
"Thou shalt not kill."
7). Seventh Commandment:
"Thou shalt not commit adultery."
8).Eighth Commandment:
"Thou shalt not steal."
9). Ninth Commandment:
"Thou shalt not bear false witness against thy neighbour.
10) Tenth Commandment:
"Thou shalt not covet thy neighbour's house, thou shalt not covet thy neighbour's wife, nor his manservant, nor his maidservant, nor his ox, nor his ass, nor any thing is thy neighbour's."

This is the law of God, which is on two stone tablets, and which is the eternal law. It applies to all residents of this planet. According to people will be judged on Judgment Day.

Verse 22. "And the LORD said unto Moses/Musa, thou shalt say unto the children of Israel, Ye have seen that I have talked with you from heaven."

Verse 23. "Ye shall not make with me gods of silver, neither shall ye make unto you gods of gold."

Verse 24. "An altar of earth thou shalt make unto me, and shalt sacrifice thereon thy burnt offerings, and thy peace offerings, thy sheep and thine oxen: in all place where I record my name I will come unto thee, and I will bless thee."

Verse 25. "And If thou wilt make me an altar of stone, thou shalt not build it of hewn stone: for if thou lift up thy tool upon it, thou hast polluted it."

Verse 26. "Neither shalt thou go up by steps unto mine altar, that thy nakedness be not discovered thereon."

CHAPTER 24:

Verse 12. "And the LORD said unto Moses/Musa Come up to me into the mount, and stay here, and I will give thee tablets of stone, and a law, and commandments which I have written; that thou mayest teach them."

Verse 15. "And Moses went up into the mountain, and the cloud covered the mountain."

Verse 16. "And the glory of the Lord abode upon mount Sinai, and there was a cloud that covered it for six days, and on the seventh day Moses / Musa shouted in front clouds. And Moses / Musa remained on the mountain for forty days and nights."

CHAPTER 31:

Verse 18. "And he gave unto Moses / Musa, when he had made an end of communing with him upon mount two tables of testimony, tables of stone, written with the finger of God ."

CHAPTER 32:

Verse 15. "And Moses / Musa turned, and went down from the mount, and the two tables of testimony were in his hand: the tables were written on both their sides; on the one side and on the other were they written."

Verse 16. "The plates were the work of God, and the writing was the writing of God, engraved upon the tablets."

CHAPTER 40:

Verse 1. "And The LORD spake unto Moses / Musa, saying, On, the first day of the first month shalt thou set up the tabernacle of the tent of the congregation. And put therein the ark of the testimony, and cover the ark with veil."

Verse 10."And thou shalt anoint the altar of the burnt offering, and all his vessels, and sanctify the altar: and it shall be an altar most holy. "Verse 20. "And he took and put the testimony into the ark, and set the staves on the ark, and put the mercy seat above upon the ark."

Verse 21. "And he brought the ark into the tabernacle, and set up the vail of the covering, and covered the ark of the testimony; as the LORD commanded Moses."

Verse 30. "And he set the laver between the tent of the congregation and the altar, and put water there, to wash withal."

Verse 31. "And Moses / Musa, and Aaron and his sons washed their hands and their feet thereat.

Verse 32. "When they went into the tent of the congregation, and when they came near unto the altar, they washed; as the LORD commanded Moses /Musa."

THE FORTH BOOK OF MOSES / MUSA: OR BOOK OF THE NUMBERS CHAPTER 14:

Verse 17. "And now, I beseech thee, let the power of my LORD be great according as thou hast spoken, saying."

Verse 18. "The LORD is longsuffering , and of great mercy, forgiving iniquity and transgression, and by no means clearing the guilty, visiting the iniquity of the fathers upon the children unto the third and fourth generation."

"And do mercy on the thousands who love and who keep my commandments."

THE FIFTH BOOK OF MOSES / MUSA: OR BOOK OF THE DEUTERONOMY

CHAPTER 10:

Verse 13. "To keep the commandments of the LORD, and his statutes, which I command thee this day for thy good?."

Verse14. "Behold, the heaven and the heaven of heavens is the LORD'S thy God, the earth also, with all that therein is, only the LORD had a delight in thy fathers to love them, and he and choose their seed, after them, even you above all people, as it is this day. So circumcise therefore the foreskin of your heart, and be no more stiff-necked. For the LORD, is above Gods and LORD of LORD'S, a great God, a mighty and terrible who does not look (who is who) or receives gifts. Gives right to the poor and widow, loves the stranger, for you were strangers in the land of Egypt. Fear your God, serve Him, and keep his charge Him. And swear by his name."

CHAPTER 16:

Verse 18. "Create yourself judges and governors in all places where I give you the LORD your God. And let them judge the people right. Don't twist the justice and do not look at who is who, do not receive gifts because the gift it spits the eyes of the wise and distorts the words of the righteous."

CHAPTER 17:

Verse 1. "Thou shalt not sacrifice unto the LORD thy God any bullock, or sheep, wherein is blemish, or what evil: for that is an abomination unto the LORD thy God."

CHAPTER 34:

Verse 1. “And Moses / Musa went up from the plains of Moab unto the mountain of Nebo to the top of Pisgah, that is over against Jericho. And the LORD shewed him all the of Gilead, unto Dan.”

Verse2. “And all Naphtali, and the land of Ephraim, and Manasseh, and all the land of Judah unto the utmost sea.”

Verse 3.”And the south, and the plain of the valley of Jericho, the city of palm trees, unto Zoar.”

Verse 4. “And the LORD said unto him, This land which I swore unto ABRAHAM / IBRAHIM, unto ISAAC, and unto JACOB / JAKUB, saying, I will give it unto thy seed: I have caused thee to see it with eyes, but thou shalt not go over thither.”

Verse 5 “And Moses / Musa, the servant of the Lord, died in the land of Moab, according to the word of the LORD. And the LORD, buried him in the valley, over against Bethpeor : but no man knoweth of his sepulchre unto this day.”

“And Moses /Musa was 120 years when he died.”

It is clear from these verses that the LORD God , he brought forth the sons of Jacob/ Jacub out of the land of Egypt the way he told Abraham / Ibrahim in the book of Existence Heaven and Earth:

THE FIRST BOOK OF MOSES / MUSA.

CHAPTER 15:

Verse 13. “And the Lord said to Abraham / Ibrahim, Know that your seed will be a stranger in a land that is not theirs and will serve them and they will torture them 400 years. But I will also judge the people they will serve and later they will come out with great treasure. And you will go to your fathers.”

Because of the great drought and the great famine, Jacob / Jacub left the country promised, there was a great famine in that land. And God said to Jacob / Jacub at night, in a vision: I' am God. God your father, do not be afraid to go to Egypt, because I will make you a great nation there. And Jacob / Jacub, went up from Beersheba, and took their cattle, and their treasure which they had received in the land of Canans. And they came to Egypt, Jacob / Jacub, and all their family. And he took his sons and his sons' sons, his daughters and daughters of their sons, and all his family with him to Egypt. And of all the souls that came with Jacob / Jacub to Egypt, except the wives of the sons of Jacob / Jacub; of them, there were 66 souls in all. And two sons of Joseph who were born to him in Egypt, all the souls that came into Egypt were 70.

The children of Israel served for 430 years in Egypt.

"By his promise made by the LORD to Abraham / Ibrahim, Isaac and Jacob / Jacub brought their sons back to the land of Canaan."

From these verses we have seen that all 12 tribes with their families came back to the land which God swore they would inherit.

600 years passed since the birth of Abraham / Ibrahim to the departure of his great-grandchildren from Egypt.

When they came out, God told them that they would divide the land they were in, by tribes and by the number of people and how many tribes there were.

The LORD appointed judges who governed the people and according to the law and judged that no one would be harmed. The judges ruled the people 400 years.

Until after 400 years they had asked for an emperor.
And the children of Israel / Jakub wished to have an emperor like the rest the peoples around them.
Until then, they had prophets and judges and judged by God's instructions and judgement in righteousness and led the people on the right path.
At that time there was the prophet Samuel whose mother asked the LORD God to give her because she had no children.
When the LORD heard her prayer she bore a son and named him Samuel. When she breastfed the child she took him to the house of the LORD in Silas and the child was still little.
And she gave the child to the LORD, as long as he was alive, he was given to the LORD and Ana, the mother of the child, and said, there is no saint as is LORD because there is no one but you and no rock as is our God.
The LORD kills and revives, descends into the grave and pulls out, can make you poor or rich, he brings you low and lifts you up.
And Samuel ministered unto the LORD in a linen vest. And the child Samuel grew before the LORD.
And Samuel grew and the LORD was with him, and none of his words fell to the ground.
And the LORD, began to appear to the child in Shiloh.
The LORD appeared to Samuel in Shiloh, the word of the LORD. And it happened everything that he said would happen to Israel. And there would be a war between the Philistines and Israel, where the Israelis couldn't win. They sent them to Shiloh to bring the ark of the covenant of God carried by the sons of Eli the priest. But they

lost the battle and the ark of the covenant of God, and they were abducted, and both were killed two sons of Eli. And the ark of God would be in the land of Philistines seven months. And they sent messengers to the inhabitants of Kirjathjearim. They brought back to the Philistines the ark of the LORD. And the ark remained for a long time at Kirjathjearim for twenty years. And all the house of Israel cried. And Samuel spake unto all the house of Israel, saying, If you do return unto the LORD, with all your hearts, put away the strange gods and turn your hearts unto the LORD, and serve him only: he will deliver you out of the hand of the Philistines."

And the children of Israel did put away Baalim and Ashtaroth, and served the LORD only.

When Samuel was old, the children of Israel asked to have a king. But Samuel was displeased when they said, "Give us the king to judge! "And Samuel prayed unto the Lord.

And the Lord said to him, "Listen to the voice of the people, what they say to you, because they have not rejected you but me, that I should not reign over them."

And so, at their request, they received the first king from the BENJAMIN tribe. King Saul.

These are the kings who reigned over Israel. Seven kings have reigned over Israel.

These are the names of the kings of Israel who reigned for 400 years over Israel:

1. Saul
2. David
3. Solomon

4. Roboam
5. Hezekiah
6. Josiah
7. Sedekia

At the time of these kings, God had his prophets who were turning away people from sin.

These are the names of the prophets of who God spoke through to the people:

Nathan, Elijah, Elisha, Isaiah, Jonah, Sophonia, Jeremiah, Avakum, Danilo, Jezekia, Zaharie, Malachia.

But the people were far from the truth and God. They sinned greatly and the lord gave them into the hands of the king of Babylon.

In the ninth year, Zedekiah king of Judah, came in the tenth month Nebuchadnezzar king of Babylon with his army to Jerusalem and pitched against it and they built forts against it round about.

And he captured the king Zedekiah, bound him, and brought him into captivity to Babylon with the rest of captive people. And they were slaves in Babylon for 70 years because of their sins and the abhorrence of the law of God.

The Book of the Prophet Daniel:

CHAPTER 9:

Verse 1. "In the first year of Darius the son of Ahasuerus the son of the Medes which was made king over the realm of the Chaldeans."

Verse 2. "In the first year of his reign, I, Daniel, understood by books the number of the years, whereof the word of the LORD, came to Jeremiah the prophet, that he would accomplish seventy years in the desolation of Jerusalem."

Verse 12. “And he confirmed his words which he had spoken against us against our judges who have judged us, by bringing upon us great evil, for under the whole heaven had not been done as had been done upon Jerusalem.”

Verse 13. “As it is written in the law of Moses, all this evil comes upon us ; yet made we not our prayer before the LORD our God, that we might turn from our iniquities and understand thy truth.”

Verse 16. “O Lord, according to all thy righteousness, I beseech thee, let thine anger and thy fury be turned away from thy city Jerusalem, thy holy mountain: because for our sins, and for the iniquities of our fathers, Jerusalem and thy people are become a reproach to all that are about us.”

Verse 19. “O LORD, hear; O LORD, forgive; O LORD, hearken and do; defer not, for thine own sake, O my God: for thy city and thy people are called by thy name.”

Verse 20. “And whiles I was speaking, and praying, and confessing, my sin add the sin of my people Israel, and presenting my supplication before the LORD, my God for the holy mountain of my God.”

Verse 21. “Yea, whiles I was speaking, in prayer, even the man Gabriel whom I had seen in the vision at the beginning, being caused to fly swiftly, touched me about the time of the evening oblation.”

Verse 22.“And he informed me, talked with me, and said, O Daniel, I am now come forth to give thee skill and the understanding.”

Verse 23. "At the beginning of thy supplications the commandment came forth, and I am come to shew thee; for thou art greatly beloved: therefore understand the matter, and consider the vision."

Verse 24. "70 weeks are determined upon thy people and upon thy holy city, to finish the transgression, and to make an end of sins, and to make reconciliation for iniquity, and to bring in everlasting righteousness, and to seal up the vision and prophecy, and to anoint the most Holy."

Verse 25. "Know therefore and understand, that from the going forth of the commandment to restore and to build Jerusalem unto the Messiah the Prince shall be seven weeks, and threescore and two weeks: the street shall be built again, and the wall, even in troublous times."

Verse 26. "And after threescore and two weeks shall Messiah be cut off, be not for himself: and the people of the prince that shall come shall destroy the city and the sanctuary; and end thereof shall be with a flood, and unto the end of the of the war desolations are determined."

Verse 27. "And he shall confirm the covenant with many for one week: and in the midst of the week he shall cause the sacrifice and the oblation to cease, and for the overspreading of abominations he shall make it desolate, even until the consummation, and that determined shall be poured upon the desolate."

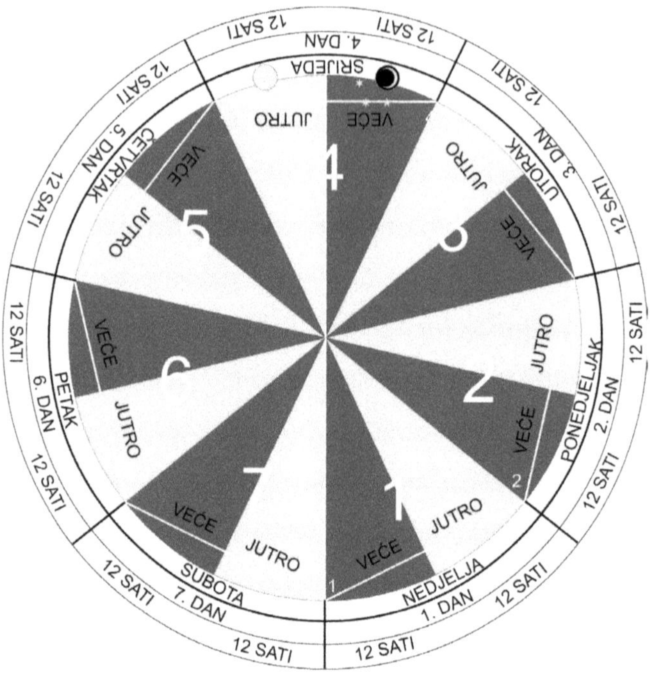

Picture 17 (WEDNESDAY MIDDLE OF THE MIDDLE)

VECE – NIGHT; JUTRO – MORNING; DAN - DAY

In this picture no 17: we see the middle of the week. It›s one week or seven days in which God created life on this Earth.

And in the middle of the week on the day, Wednesday , exactly on that day he created the Sun, and the Moon, and all the army of Heaven.

And exactly in that place; is the Sun; on the middle of the day is (noon) sixth hour of the day. At that point, in the middle of the day, the truth, would be divided from lies, and the light, would be divided from darkness and the death, would be divided from the life.

And he confirmed the covenant with many for one week: and in the midst of the week he caused the sacrifice and the oblation to cease for the overspreading of abominations he shall make it desolate, even until the consummation, and that determined shall be poured upon the desolate."

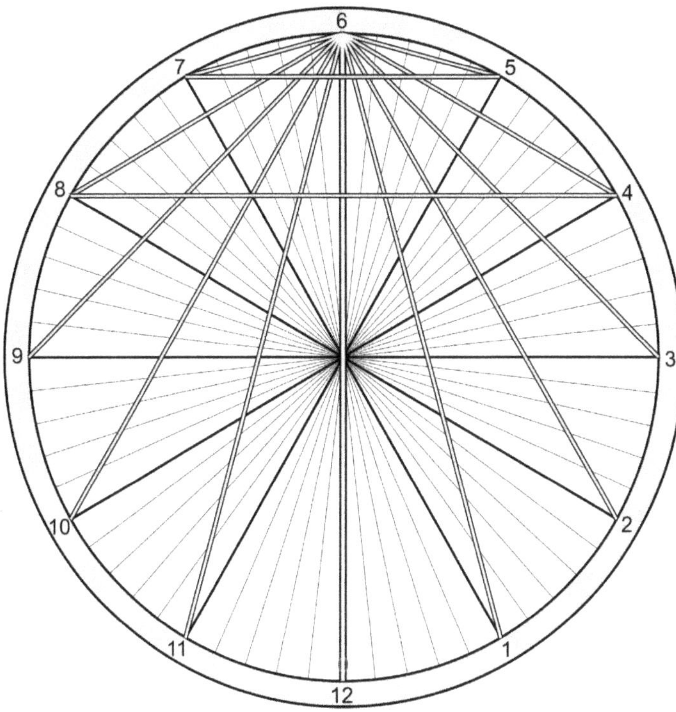

Picture 18 (CENTER of all CENTER).

By the determined end it will be poured out on the desolate exactly where the centre happened of all the events that the lines of this earth are referring to.

And these are: the line of light or the line of truth, because in light, in the day nothing is hidden.

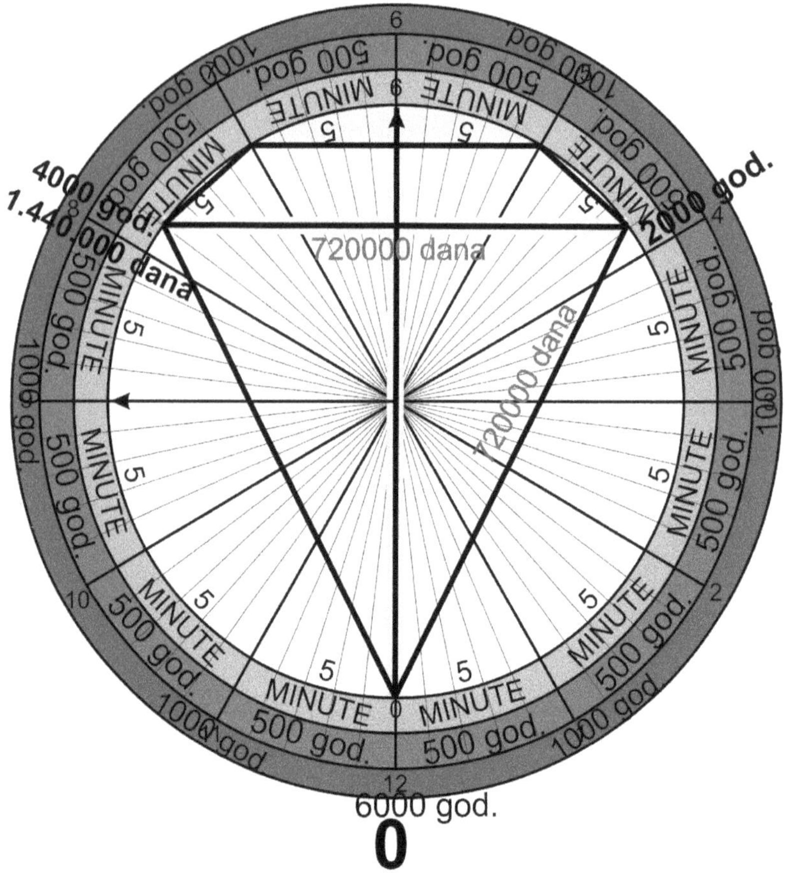

Picture 19 (STONE JASPIS / ALEM IN TIME AND SPACE).

The line of time that takes us from the first second of creation which brought us to exactly 4.000 years since time is counted on this earth.

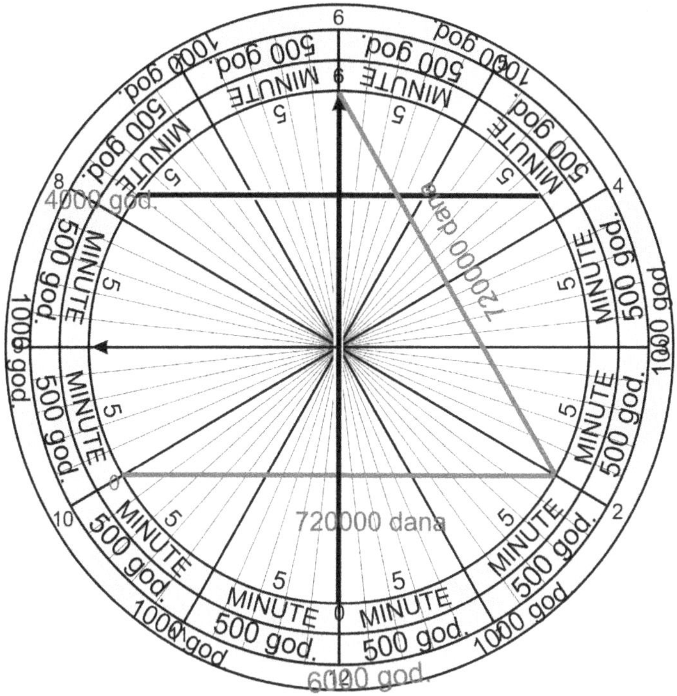

Picture 20 (BLOOD LINE IN TWO TIMES).

It's from when Adam / Adem was kicked out from paradise. Because of disobedience to the law of God and because of sin, he had to leave it; God told him that day, when he tastes from the forbidden tree he will die. The penalty for sin is death. And Adam / Adem would be cast out of paradise. But on the same day he did not die because the Creator out of his love for his firstborn son, found a replacement for his death and his sin was redeemed. Only this way was it possible to save the human race from eternal death. The blood line, which leads us from Adam / Adem's imperial blood because Adam is created as the emperor of this planet.

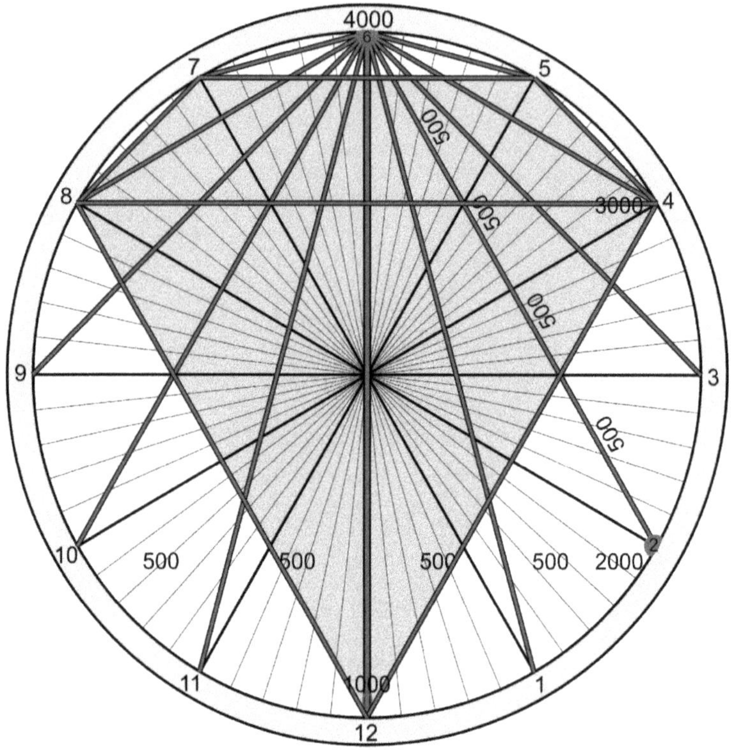

Picture 21 (SECOND CHANCE IN ALFA AND OMEGA).

The line of grace and love that was created in God's love and in the plan of salvation. In which the father of heaven gave men another chance, to receive forgiveness for sins through the blood of the Messiah, and to return there where they were created, into the Garden of Eden, Paradise or Jennet, into eternal life. Everyone has free will and every individual can choose, by his own free will, in which direction he will go through his life or through his time given to him by the creator at the time he was born.

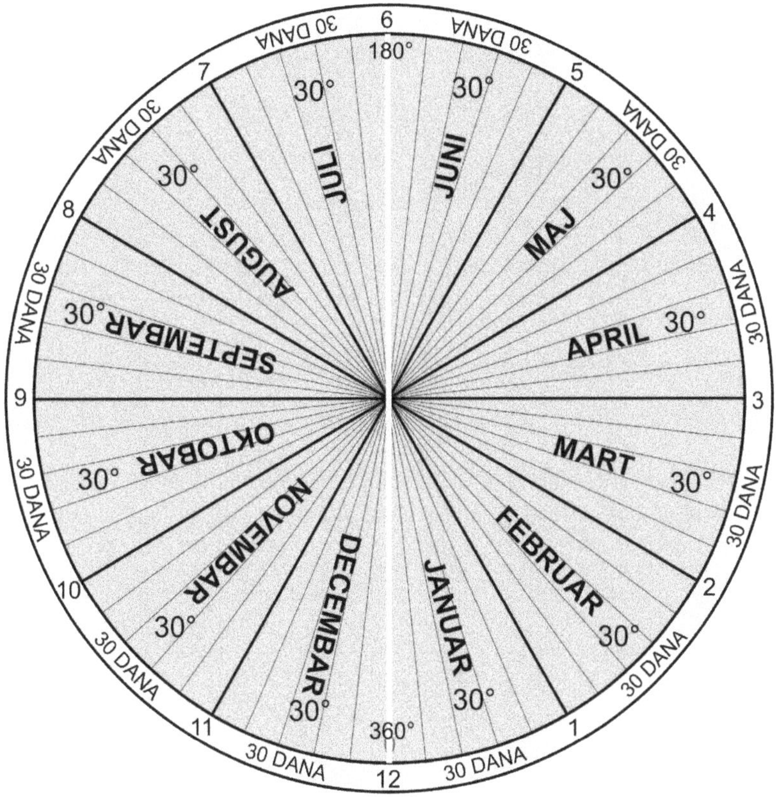

Picture 22 (12 MONTHS IN ITS OWN - ALPHA AND OMEGA).

Shows the Alpha and Omega of one year.

It shows (hourly) time, 12 months beginning as well clock, from right to left

1st. Month is January

2nd. Month is February

3rd. Month is March

4th. Month is April

5th.Month is May

6th. Month is June

7th. Month is July

8th. Month is August

9th. Month is September

10th. Month is October

11th. Month is November

12th. Month is December

In this picture that shows one year, we also see that half the year right in the middle of the annual calendar. And yes light passed from 0 time to the 6th the month, of June, which is in the middle of the calendar for one year. There fits the wooden cross on which the plan of salvation for the sinful man is fulfilled.

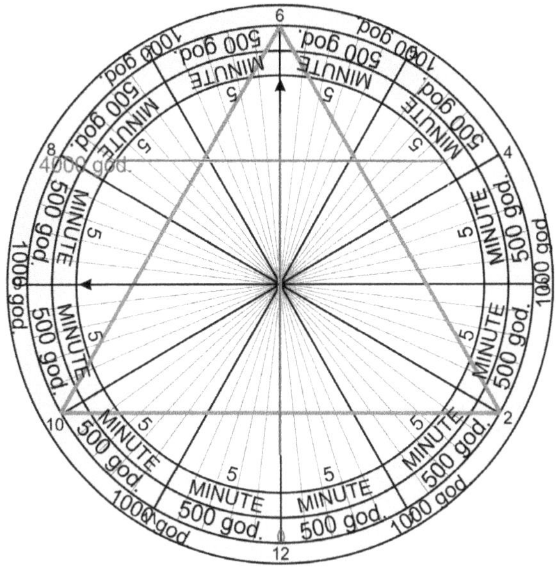

Picture 23 (TIME IN THREE DIMENSIONS).

Indicates an hour that beats a second, 1 minute, 1 hour, 1 day, 1 year, 500 years, 1.000 years.

It then shows all the 6.000 years that has already went through this planet and which began to count from the beginning of sin and from the beginning of Adam / Adem's life as a sinner.

Because of his conscious disobedience to God, he obeyed his wife and lost his eternal life. Because his wife was deceived by a snake ; but he was not deceived by the serpent, he knowingly violated God's commandment. Thus the sin of one man entered into all men. Their own sinful genes transmitted to their own children. Although he lost eternal life, Adam / Adem was left with his pure, imperial blood which he had in himself.

It was to be transmitted until the time specified by God. Exactly in a particular year, day, hour. The redeemer / saviour needed to be born who was Adam / Adem's replacement for his death and who was a replacement for the death of all people who believe in him and through the grace of God receive forgiveness of inherited / acquired sin. To return back again to the place that we had in the first day of creation at Garden of Eden.

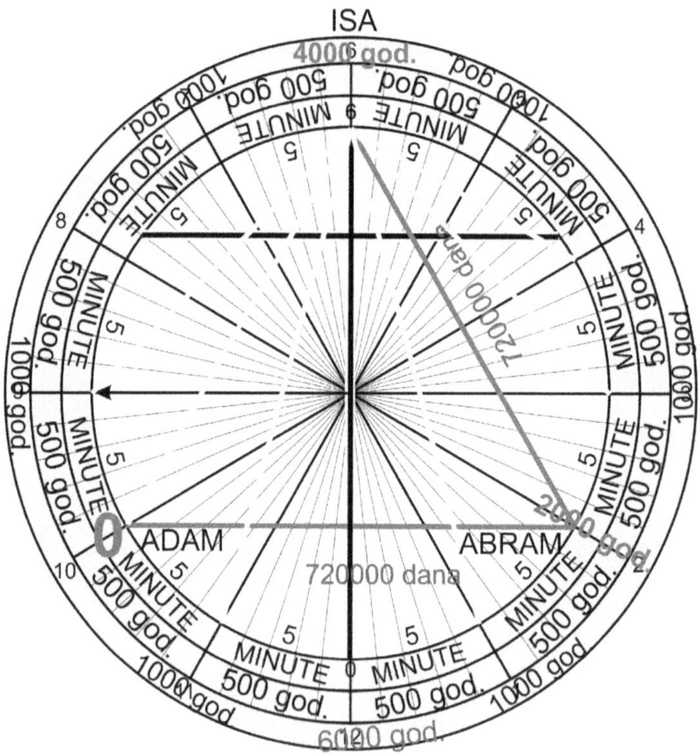

Picture 24 (DAYS AND YEARS THROUGH THE WEEK).

It shows with a red line, which goes from Friday, the sixth day in which God created man and woman.

On the 10th. hour of the day, there is also the sixth day of creation; Friday and also the last day of creation.
Like the 144th. hour at which the creator finished creating life on this earth. We see a red line showing, at the same time a blood line which needed pure Imperial blood to carry to birth of the anointed Messiah. He should have been born of that pure blood.
And his name is KING over KING and MASTER over MASTER .
The line goes straight to 4.000 years, through firstborn sons, who were determined, who respected and believed in one God who created everything visible and invisible and only worshiped and believed in him.
So holy blood through Adam / Adem's son Seth; would transfer to Noah / Nuh, over the flood and over Noah / Nuh's son; to Abraham / Ibrahim.
From Adam / Adem to Noah / Nuhh 1.000 years passed:
From Noah / Nuhh to Abraham / Ibrahim – 1.000 years passed:
From Abraham / Ibrahim to Emperor David / David – 1.000 years passed:
From Emperor David / David to Jesus / Isaa / Messiah – 1.000 years passed:
And exactly 4.000 years (Old Ages) were fulfilled. And that is exactly 4.000 years where Jesus / Messiah was born.
There is where it was fulfilled with 1.440.000 days since the beginning: of Adam / Adem sinful life.
At that place, at 4.000 years after sin occurred.
The firstborn and first created son of God in heaven was born on Earth to the will of his father.

And, he would be conceived by the Holy Spirit without a father on Earth; and Mary / Mejrema ; bore him in Bethlehem; After his death he returned to Heaven to his father.

He came and replaced with his death the death of the firstborn and the first created - on Earth the son of God, Adam / Adem.

And with his holy life and his holy blood he redeemed the sinner's that no one who believes in him should perish but have eternal life.

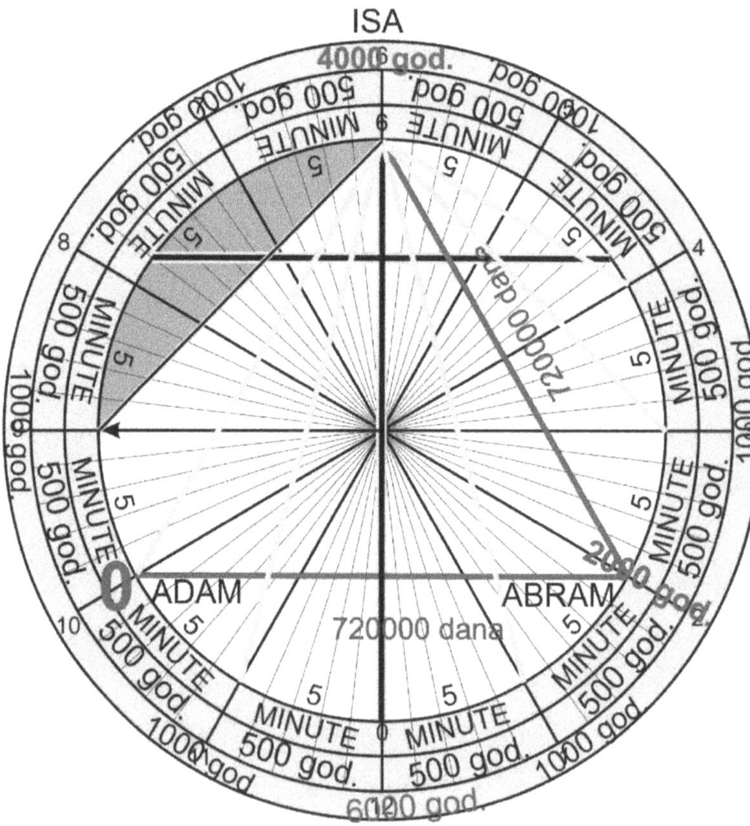

Picture 25 (COMPLETED RESCUE PLAN).

Shows the time, day, and time of fulfilment of the plan of salvation and redeemed from the sin of Adam.

The creator himself confirms and says who created us, redeemed us and who will raise us from the dead.

On Judgment Day, in which we will all be judged; some for life eternal, and others to shame and death eternal.

In this figure No.25

We see a time of 6.000 years, where it is seen that the 4.000 years have long passed.

That the plan of salvation was long ago fulfilled, and that people from this earth already had their time.

Exactly the Universe: and God, through the Universe and time, gave to us a signpost and a direction that we should not wander and stray and lose what matters most , the notion that everything created has its time.

And time brings and takes everything created as well as that time also has its time.

The time is passing and it almost past. That means the figure shows the time that has passed on this Earth.

From the events described above, confirmed by the Universe through the light that illuminated time, that we may follow time, and that time has revealed and confirmed the truth of the events that have happened. Through time we can see the events that have been behind us and which is currently in front of us. And to be able to understand the messages through the images of the Universe.

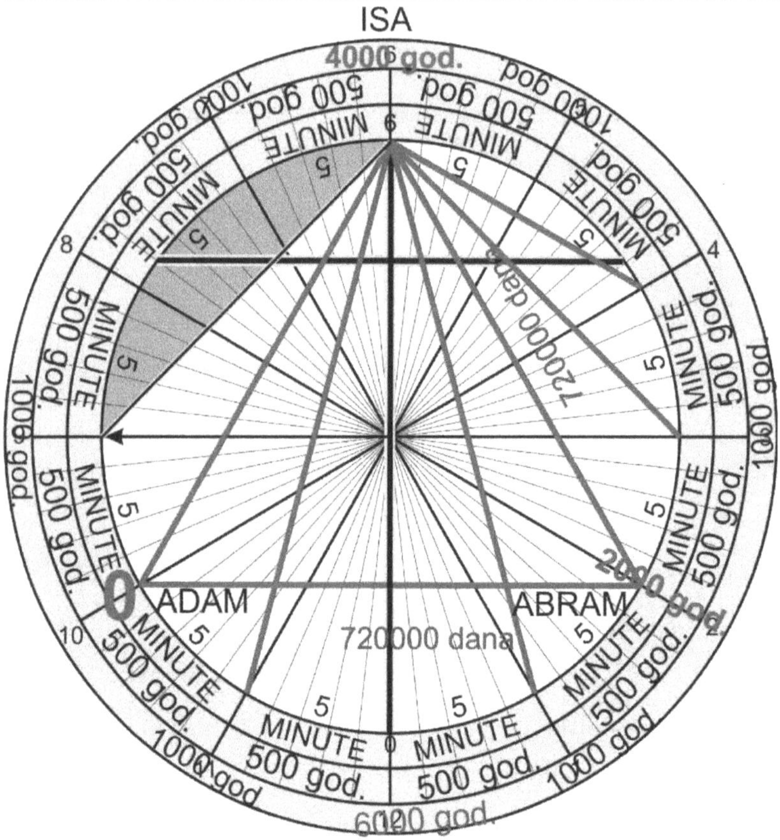

Picture 25a (COMPLETED SALVATION PLAN IN DAY AND HOUR)

From this image no.25a.

We see the message, the clock stands at Number (6) and Number (9). And the black line goes from No.(12) to No. (6) upright and horizontal from No.(8) to No.(4).

These lines show the cross-tree on which Jesus / Isaa died. Then red lines indicate that he with his blood, with his love and by his grace he embraced the whole planet.

And that he gave his life for all nations, tribes and languages that none perishes who believes him, but to have eternal life.

The hand standing at the sixth hour of the day (6) and the ninth hour of the day (9), shows the time when Jesus / Isaa was dying on the cross.

In the Gospel of Luke:

Chapter 23: Verse No.44

And it was at about six o'clock, and there was darkness in all the earth, until the ninth o'clock and the Sun was darkened, and the curtain of the church was cut in half. And Jesus with a loud voice, said ;

Father, into thy hands commend my spirit; and saying he exhaled.

The full glory of the ascension was shielded by a veil.

For the apostles it was the most precious fact that Christ; ascended to heaven in tangible form as their divine teacher. The image of the creator as a compassionate friend and glorified redeemer was the last impression left before the student's eyes. Moses / Musa, covered his face to obscure the glory of the law which reflected on him. So is the glory of Christ, his ascension was shielded from human eyes. The eyes of mortals could not see the glow of celestial accompaniment opening the magnificent doors of the sky in welcome to the son of God

Christ's departure to heaven was revealed to his disciples and indescribable glory, their eyes could not stand it. That they saw innumerable multitudes of angels and heard thunderous outbursts of triumph with the heavenly openings, when the eternal door opened, the opposite between this glory and their torturous life in this world would be so great that they would not be able to bear the burden of this worldly life. And they would not be willing to bravely and faithfully execute the order the Saviour gave them. Not even a comforter, a holy spirit, which was promised to them would not be properly understood, neither would their hearts bear the shame of the mockery, and death, if needed. Their senses should not have been so enthused by the glory of heaven that they lose sight of the character of Christ manifested on Earth, which is supposed to be reflected in each of them. Before their eyes, the beauty and exaltation of his life should have been the perfect harmony of his properties and the mysterious unity of God and man in his nature. His apparent departure from this world was consistent with the meekness and gentleness of his life. In his ascension to heaven, Christ also brought his dedicated human nature. He also brought his human nature to heavenly courts and carriers throughout eternity as the one who redeemed every human being found in the kingdom of God

Christ came into the world as a saviour. He represented God on Earth. As a personal saviour, he ascended into heaven and as he left, so will he come again, as a personal saviour.

Christ wanted to be in a position from which, through few and simple tools, he would be able to do the most significant work.

The plan of redemption is comprehensive; but those who participate in his realization is not many. And they all depend on one another while they all work together in the utmost simplicity and complete harmony. The holy spirit is the representative of

Christ, and when that spirit is appreciated, when they, whom he controls, pass on to others the energy with which they are perpetuated, it touches the invisible wire through which the electricity is transferred to everything. Oh, if everyone could understand how borderless divine sources are.

The holy spirit gives divine authority. You will be received by force when the holy spirit descends upon you. And the Holy Spirit gives divine authority. "But they will receive you: when the holy spirit descends upon you. Christ / Messiah says." And you will be my witnesses. The Holy Spirit will unite with the testimony of living witnesses to issue the world that reminder. The worker for God is the tool through which heavenly confessions are made. And the Holy Spirit gives divine authority to the words which he conveys.

The treasures of the celestial powers are not locked. When, 50 days was over, together all the disciples, the apostles, as one. And all of a sudden, there came a howl from heaven like a great wind, and filled the house where the students sat. And they were shown divided tongues like fire. And they were all filled with the holy spirit, and began to speak in other languages, as the spirit gave them, they said, "God is willing to give the same blessing to us if we seek him earnestly."

And after this John said, " I saw another Angel coming down from heaven, which had a great region, and the Earth shone with the glory of it . Then, as on the day of 50 days people will listen to the truth that will be presented to everyone in his language. All those who want to sincerely serve him, God can breathe new life."

The seeds sown by the greatest teacher of the world could not remain unknown. For three and a half years the Son of God was in Judea explaining the Gospel message of truth and working with great signs and miracles. The seeds were sown after his ascension a great harvest ensued.

After one sermon more souls were converted than during all the years of Christs public services. So god will act when people obey the control of the holy spirit

The disciples began to realize that Christ / the Messiah had regained his glory and that he was now their representative in the heavenly courts. They could speak with freedom because the holy spirit was putting words in their mouth.

There is no excuse for staying ignorant.

Jesus / Isaa, when he was in torment, said. "Forgive them, Father, they do not know what they are doing. I know out of ignorance they are doing this."

But this ignorance did not constitute an excuse for the action being done because they were offered a great light.

It is said that they would not crucify Christ if they knew that he was the prince of life.

But why didn't they know that?

Because they chose to remain ignorant. They were not interested to study and study the prophecies and it turned out to be that their ignorance led to eternal destruction.

They had the strongest evidence on which they could base their faith, and before God they were obliged to accept these evidence that he gave them. Their disbelief made them guilty of the only begotten son of god.

What does God say? We do not need to ask what the practice of people is or what the customs of the world are. We don't have to ask what to do to win the approval of the people or how to secure the tolerance of the world.

The question of most urgent interest to every soul is, what does God say.

We need to read, the word of God, and obey it without departing not a single title from its demands, regardless of human traditions jurisdictions often referred to.

Repentance is the gift of Christ/ Messiah, just like forgiveness, and it cannot find a place in the heart in which Christ has not done his work. As you are without Christ/Messiah, we cannot receive forgiveness, so we cannot repent unless his spirit awakens our conscience.

Christ / Messiah attracts the sinner with his love expressed on the cross, and it stirs the heart, impresses the mind and inspires a longer humility and repentance.

Circumcision has no value after Christ / The Messiah's crucifixion and those who adhere to faithful word in science, to be able to explain with healthy science to those who oppose. There are many disobedient, empty-spoken and mind-scammed, especially those who need to close their mouths, who have entire homes who they learn the wrong thing for the sake of dirty gains.

The testimony is true, for the sake of that cause, drive them without austerity, to be healthy in faith, not listening to the Jewish witchcraft or the commandments of people who deter from the truth.

In Paul's days, there were people who kept lingering on the question of circumcision by presenting evidence from the Bible, the Torah, the Tevrata which have affirmed the obligation for the Jews, but the teaching became insignificant in Paul's time.

Christ / Messiah had already died on the cross and the circumcision of the body no longer had any value. Both symbolic service and ceremonies related to it were abolished

The one who represented the true (non-symbolic) lamb of God, was already sacrificed for the sinful man and the shadow is lost in the reality.

The gospel is the only one and the truth is only one. It is incorrect of the day, which is so often presented, that there is a contradiction between the Old and New Testaments, the Law of God and Christ / Messiah the gospel. It demands what is required of the Jews and of what is required after Christ / Messiah.

Every soul saved in the Old testament period was saved by faith, merit, Christ / Messiah. Just as we are being saved today too. The patriarchs and prophets saved themselves by believing in the promised Messiah through the sacrificial lamb, which represented the living a sacrifice to replace the lamb with his life and his blood has been slaughtered for 4.000 years, and instructed people to understand that someone bigger than that lamb, has to pay for the sin of Adam / Adem and to redeem all mankind from inherited sin. Yes anyone who by faith seeks forgiveness for his own sins, as well as for hereditary ones, dies not to eternal death, but through the already paid, our sin, through Christ / Messiah also receives eternal life. The promise of the gospel has been given to Adam / Adem and Eve in the Garden of Eden, when by their transgression they separated themselves from God.

God is constantly operating in nature. Nature, like his servant, does by his will. Nature in its action witnesses to the presence and activity of a reasonable being who manages and maintains all in accordance with his will.

Its continuous movement around the sun and its treasures to which it maintains as much humanity, the Earth does not bring to some force that is not insanely related to nature itself. The hand of infinite force continuously maintains this planet.

The force that maintains it at all times in its eternal orbit around the Sun is just a force of God. The God of Heaven is constantly at work. With his force, vegetation flourishes, each leaf develops and opens every flower.

Continuous heart rate and breathing are not the product of a mechanism that once put on the move continues its work all the time. Every breath of the heartbeat represents a continuous testimony of the force of the always present God. In God we live, move and exist. Each breath, every heartbeat is a constant testimony of the power of the ever present God.

God created the Sun and gives it to be born every morning and shines on in the firmament. He opens the constitutions of heaven and gives rain. He makes the grass grow, he gives snow like the wool, pours snow like ashes. He releases lightning with rain and brings the wind out of his stable. Although he finished his works in the creation of the world he uses everything he created like his servants for the good of a man.

Christ / Messiah say's. "My father works and now I work too."

The Book of the Prophet Daniel:

CHAPTER 12:

Verse 1."And at that time shall Michael, stand up, the great prince which standeth for the children of thy people. And there shall be a time of trouble, such as never was since there was a nation even to that same time: and at that time thy people shall be delivered, every one that shall be found written in the book."

Verse 2. "And many of them that sleep in the dust of the Earth shall awake, some to everlasting life, and some to shame and everlasting contempt."

Verse 3."And they that be wise shall shine as the brightness of the firmament; and they that turn many to righteousness as the stars for ever and ever."

Verse 4. "And you, Daniel, shut up these words and seal this book even to the time of the end. Many will run to and fro, and knowledge shall be increased."

Verse 6. "And one said unto a man clothed with linen, which stood upon the waters of the river: How long shall it be to the end of these wonders?."

Verse 7. "And I heard the man clothed in linen, which was upon the waters of the river, when he held up his right hand and his left hand unto heaven, and sware by him that liveth for ever that it shall be for a time, times, and an half; and when he shall have accomplished to scatter the power of the holy people, all

these things shall be finished."
Verse 8. "And I heard, but I understand not: then said I, O my LORD, what shall be the end of these things?."
verse 9. "And he said, Go thy way Daniel: for the words are closed up and sealed till the time of the end."
Verse 10. "Many shall be purified, and made white, and tried; but the wicked shall do wickedly: and none of the wicked shall understand; but the wise shall understand."
Verse 11."And from the time that the daily sacrifice shall be taken away, and the abomination that maketh desolate set up, there shall be a thousand two hundred and ninety days."
Verse 12. " Blessed is he that waiteth, and cometh to the thousand three hundred and thirty – five days."
Verse 13. " But go thou thy way till the end be; for thou shalt rest, and stand in thy lot at the end of the days."
Year after year passed, and the representatives of the people of God began teach their people that a Messiah will come who will overthrow the Roman authorities and liberate the Jewish people from the occupiers.
No prophet has ever prophesied such a Messiah.
About 600 years before Messiah was born, the prophet Isaiah; writes of the Messiah: He was despised and rejected among men, a man of sorrows and acquainted with grief and we hid as it were our faces from him. He was despised and we esteemed him not. He had borne our griefs, and carried our sorrows yet we did esteem him stricken, smitten of God, and afflicted and we thought that he was wounded that God would beat him and

torture him. All of us as sheep have gone astray, we have turned Every one to his own way and the LORD had laid on him the iniquity of us all. He was oppressed and he was afflicted, yet he opened not his mouth, he is brought as a lamb to the slaughter, and as a sheep before the shearers is dumb so he opened not his mouth. Isaiah (53)."

Jesus / Isaa said to Pilate, Roman Judge: "My kingdom is not of this world."

When the prophecy of the Messiah began to be fulfilled the Jewish nation did not want to recognize it because he did not come to them as they had expected it . We can see how the leaders of the people even before Christ / The Messiah's was born, they misinterpreted the prophecy about Messiah before his first coming. And won't it be the same before the Messiah's second coming in glory?

The Gospel of John:

Chapter 8:

Verse 25. "Then said they unto him, Who art thou? And Jesus said unto them, Even the same that I said unto you from the beginning."

But the people of that time, like now, did not have time to study the scriptures and they gave the information how they wanted it to be. That is also how they thought the people.

At that time, the Roman Empire ruled over them and their desire was for God to free them from Roman slavery.

The Messiah / Christ came to fulfill the plan of salvation because he offered to God his sinless life, for a sinful and fallen man. That no one who believeth in him should perish but, have eternal life.

But Jesus / Isaa , said to Pilate, the Roman judge: "My kingdom is not from this world."

"The Lord said to the first married couple in the Garden of Eden: "Because the day you taste from it, you will die."

"Why did Adam / Adem and Eve / not die on that day when they broke the law of God? ."

Adam/Adem and Eve / Hava; they were kicked out from paradise, but at the gates of paradise they immediately offered sacrifice for sin.

And looking at the sacrifice, they believed that when the Messiah came he would be a sacrifice that will purify everyone's sins with his blood.

From then until the Messiah's death, the people of God offered sacrifices over which they confessed every sin for the transgression of the God's Low.

In:

The Gospel of JOHN, or the book name, INDIR:

CHAPTER 1:

Verse 29. "The next day John seeth Jesus coming unto him, and saith, Behold, the Lamb of God, which taketh away the sin of the world."

The apostle Paul writes to us: "And almost everything is purified by blood, according to the law and without shedding of blood there is no forgiveness. "The LORD said to Moses / Musi, to say to the sons of Jacob / Jakub: "For the soul of the body is in the blood, and I have appointed it for you, for the altar to purify souls, yours because blood is what cleanses the soul."

Thus, only through the Messiah's blood shed on the cross the sinner can-will receive the forgiveness for their sins. No one else can forgive the transgression of the law of God except the one who died on the cross of Golgotha.
None of the people died for the sins of the world, which is why none of the people on Earth can erase the transgression of God's law.
Isaa / Jesus / Messiah says about himself, "For I came down from heaven, not to do my own will, but the will of the Father who sent me. "It says to us here, that before Isaa / Jesus/ Messiah came down from heaven, lived in heaven. Is this not quite clear enough? That the beginning of Christ / Messiah is not with being born on this earth.
After 6 days Jesus / Isaa, took Peter, James and John and brought them out to the high mountain alone. He transformed in front of them and shone his face as a Sun, and their garments became like snow.
Jesus / Isaa resurrected Lazarus and he says. "I am the resurrection and the life, whoever believes in me and if he dies, he will live."
In: The Gospel of John:
Chapter 11:
Verse 26. "Now from this information which is extracted from the holy books from the Old Testament, or the Torah, or the Tevrat, we have received important information that tells us which direction to follow that goes further towards God and eternal life."
Isaa / Jesus / Messiah will come a second time, as promised by Isaa / Jesus to all the disciples, when he ascended into heaven, he would come again. When they looked at him as he when it

ascended to heaven; behold, two men stood before them in white robes and said to them, “Men, Galileans, why are you looking at the sky? This Jesus / Isaa, who was taken from you to heaven so it will come as you see it going to heaven.”

An explanation on the written text: Which all up to this point, walked in a line or direction of light, direction of time, direction truth and bloodline direction, all these directions lead us to one place and show only one important piece of information:

That the direction of light goes from the first second that went from zero, from the moment when God the creator of everything visible and invisible of the divine first in the first second separated the darkness from the light, and separated the truth from lies and through the light in which he was showed us further direction of movement towards the Garden of Eden or life.

That line of light guided us through 4.000 years and brought us to the place where it should have opened our eyes and shown the next direction.

This is the direction in which God separated darkness from light. And light passes through space, where it is shown to us that God created the Sun on Wednesday, i.e. right in the middle of the week and its light in which he created day and night until Wednesday, and divided the water into salty and sweet and created the sky and the sea, pulled the Earth out of the sea and on it created life. Then on Wednesday God created the Sun, the Moon and all the host of heaven.

On Wednesday, in the middle of the week, the Sun continued to maintain the light which took over a direct line of light from

God and that means, light + Wednesday + fourth day of the week (Wednesday) = 4. 000 years.

This is where the most important event for the human race took place.

At this point time and light are combined into one whole, which in translation means: truth- light + time. These directions lines confirm the same thing and this is true at 4.000 years from the beginning of Adam / Adem; sin and his life after leaving paradise God's plan of salvation for Adam and Eve as well as for their descendants began to be fulfilled immediately.

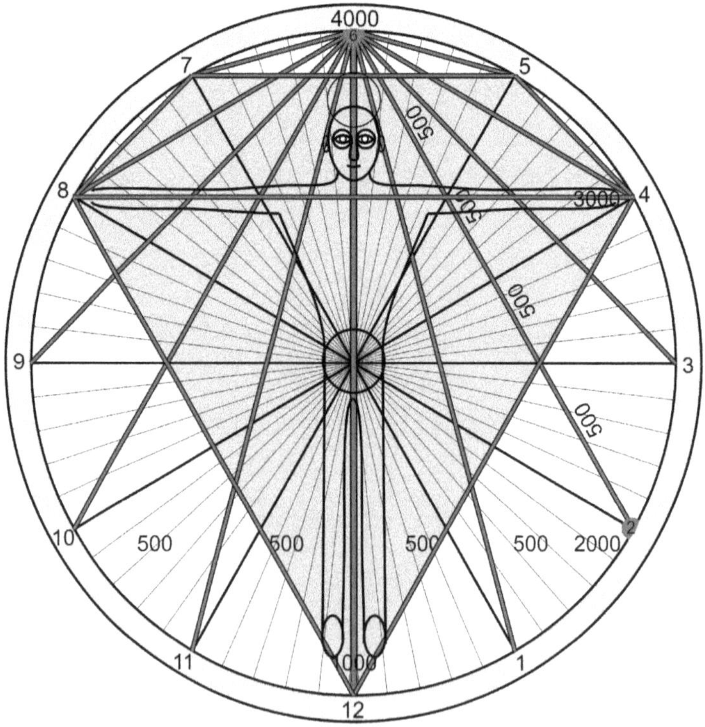

Picture 25b (CHRIST-MESSIAH / STONE-JASPIS).

It shows the light and time that has been counted on this Earth since the moment of creating the first second, the first minute, the first hour, the first day and the first week in the information we have in this book.
The holy books describe exactly how the Earth was created and the life on it and in what order is it created life. Immediately, in the first second, God created light and in the light, time, and that immediately began to count; time, and which is still being counted today.
Then the light brings us into the line of time, and the line of time also leads in the same direction as the line of light. It just leads us through time because all these lines have the same direction to confirm truth through its directions.
The line of time begins, also, from the first second of which God started counting time in the first week of creation.
The creator has given us all the information to be used in counting time
From the information provided so far, we can see that one week has 7 days, 6 working days and one holy day 7thday, it is Saturday.

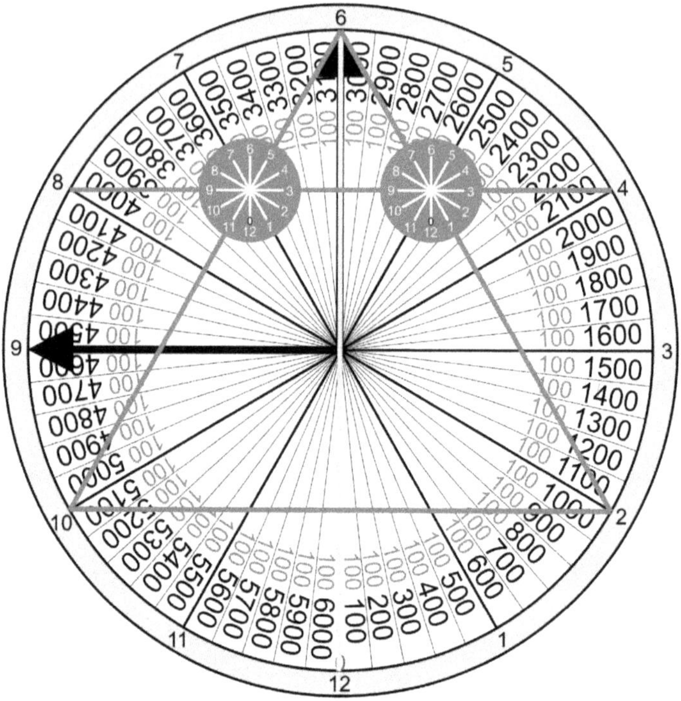

Picture 26 (TIME IN ITS ALFA AND OMEGA).

The time from the first second to the last, which will be when The Creator, as he created light and life and at a certain time extinguishes that light and life on this Earth. We see the light pass from the first second to the middle of the day, which is at the 6th hour and noon is (not at the 12th hour) of day because 6 o'clock is the middle the day:

Days are counted from Sunday:

Day 1: is Sunday

Day 2: is Monday:

Day 3: is Tuesday:

Day 4: is Wednesday:

Day 5: is Thursday:

Day 6: is Friday:

Day 7: is Saturday:

Since days are counted as time, from right to left, Sunday is the first day of the week.

Since the days are numbered : and would be evening and would be morning day one or day two we came to the information that every night and every day are counting down their hours.

And every night has 12 hours, and every day has 12 hours, that's where the information is that noon is the middle of the day, is at 6th o'clock, and that the middle of the night is at 6th o'clock, in the middle of the night. Here Wednesday fits in the middle of the week . On Wednesday it is also the middle of it all. As the middle of the day, the middle of the month, the middle of the year, and the middle of 6.000 thousand years.

Wednesday as it also got its name from the creator has its own line along which we need to get important information. Such as truth, time and space. In the sixth day, which is Friday, in which God created man in his form and right at the 144th hour, of life on this earth.

The creation of the first week, from that day to the present are counted. Each week has six working days and one holy day. All six working days in which God created life on this Earth are days: Sunday, Monday, Tuesday, Wednesday, Thursday, Friday. That are creation days or workdays.

And on the seventh day, which is the Sabbath, he created nothing more.

Saturday is counted at 24 holy hours, so one week in total has 168 hours.

Since the holy hours have their own sacred direction, we summarize them separately, so we will get this information:

The first working week, has six working days, and in six days there is 144 working hours.

10. working weeks have 60 days, and they have 1.440 working hours.

100. working weeks have 600 days, and they have 14.400 working, hours.

1.000 working weeks have 6.000 days, and they have 144.000 working, hours.

10.000 working weeks have 60.000 days, and they have 1.440.000 working hours.

Holy Hours Saturday:

1 Saturday in the first week of creation has; 1 day and it has 24 holy hours.

10 Saturdays in 10 weeks has 10 days and they have 240 holy hours.

1.00 Saturdays in 100 weeks has 100 days, and they have 2.400 holy hours.

1.000 Saturdays in 1000 weeks has 1000 days, and they have 24.000 holy hours.

10.000 Saturdays in 10.000 weeks has 10.000 days, and they have 240.000 holy hours.

When we multiply the first Saturday and its 12hour . (Nights) + 12hour.(days) by the minutes that were in that day, we would get 24hour. x 60 minutes = 1. 440 minutes. These are the minutes of the first holy hours that immediately began to count separately from working hours.

When we turn 240.000 holy, hours into minutes, we will get:

We will get 240.000 x 60 = 14.400.000.

And that 240.000 hours have = 14.400.000 minutes.

But when 240.000 holy hours divide by days it is :

240.000 : 24 = 10.000 days.

10.000 days divided into years, (at 360 days a year).

10.000: 360 = 27.777777 We see here that it is exactly 27 years. 7 x 7 numbers continue to one whole line.

It shows the year, day and time when Jesus / Isaa began his evangelical work. When he began to fulfil the plan of salvation for which he came to this Earth.

When 3.5 years of Christ / Messiah work on this earth is added to this sum (3.5 is a number located in the middle of the week or at Wednesday) we get the information that 3.5 years lasted Christ / Messiah mission in his working hours (3 years x 360 days = 1.080 days +1/2 years = 180 days = 1.260 days).

When we add up these days to the days in which he began to fulfill his plan of salvation it adds that he completed his work at 31 BC. (27.77778) year. This number shows the exact time when Jesus / Isaa started his work and the day and time when he finished his work on this Earth.

It is also the day and hour of the birth of Christ and his death. Everything is shown to us by the Creator as a confirmation of the truth that everything that exists it tells us instructions on who is and what is the only direction through which we can get back to God from which we are separated because of our inherited sins. All this information leads to Golgotha on the cross and to the one who was on it where the plan of salvation is fulfilled through all the perfect Alpha and Omega and that through him God created a bridge over which eternal death can be bridged and eternal life entered.

When we multiply 10.000 weeks, by the hours in them we will get 1.440.000 hours.

Then, when we divide 60.000 days into 360 days, how much 1 year has we get 166.666.667, and multiply by 6 = 1000 years.

Or 60.000 days x 6 = 360.000 days, has in 1000 years.

Or, one year is 360 days x 1000 years = 360.000, days in 1000 years.

It revealed who was born and for whom the entire Universe showed and indicated who was born at exactly on 4.000 years.

And that's 360.000 days in 1000 years x 4.000 years = 1.440.000 days. That number confirms the time of Old Era, the time which confirms who was born that year, that day, that minute, counting in working hours or holy hours. We always get a confirmation that only one person was born and fulfilled the plan of salvation, so sinners from this Earth return to eternal life again and return to the one who created them, who redeemed them and who will raise them from the dead when the time is right.

24 holy hours x 10.000 weeks = 240.000 x 6 = 1.440.000: they confirm the same number of 240.000 holy hours there are in 10.000 weeks.

We divide 10.000 holy days into one year or 360 days – 10.000 : 360 = 27.777.778

The timeline leads to seconds, hours, days, months, years and 1.000 years.

1 minute has 60 seconds.

6 minutes x 60 seconds = 360 seconds.

1 hour has 60 minutes.

6 hours x 60minutes = 360 minutes.

1 day has 12 hours.

30 days has 360 hours,(days) or (light).

30 nights has 360 hours of night or darkness. 1 month has 360 hours a day and 360 hours a night.

1 year has 12 months x 30 days = 360 days.

10 years has 3.600 days.

1.00 years has 36.000 days.

1.000 years has 360.000 days.

This is where the timeline brought us to the same time, just the time at years that fits in a minute, a second, a day, a year, right at 4.000 thousand years.

1000 years x 360 days = 360.000 days.

360.000 days x 4.000 years = 1.440.000 days.

On the fourth day of the week, which is Wednesday or midweek, in the place where God created the Sun, the line of Time put together all events. It is 4.000 years since life became lost after paradise was lost.

That's where meets 1.440.000: day of sinful life on this planet.

It meets 60.000 weeks with 60.000 x 144 hours = 8.640.000 working hours.

We divide 8.640.000 work hours into days, and the day has 24 hours, that is = 360.000 days.

1.000 years has 360.000 days.

360.000 days in 1000 years, multiplied by 4, how many there are in 4.000 years is also 1.440.000 days.

Holy Hours lead this line - in 60.000 weeks;

60.000 Saturday or 60.000 holy days.

60.000 x 24 holy hours = 1.440.000; and that is midweek or Wednesday.

If it is counted by God: "And it would be evening, and it would be morning," the day has, 12 hours and his middle is at the sixth hour.

Mid-year, twelve months; 12 : 2 = 6; 6 months.

Middle of 6.000 years; 6000 : 2 = 3.000 years.

That's 1.440.000 (days), of the 4.000 years, on which, Jesus / Messiah fulfilled the plan of salvation.

And on the, (1.440.000) (Holy hour) at which Jesus / Messiah fulfilled the plan of salvation.

And there it is, exactly at 4.000 years , as well the fourth day of the first week of creation, on that fourth day on Wednesday, God created, the Sun, and all the host of Heaven.

At that point in 6th, hour in the middle of the day, the redeemer of sin who came to overcome sin and Satan, in that place, where Adam / Adem , was created at 6th. day on Friday at, 144 the hour, of creation.

Right here, in the same place, on the sixth day, on Friday, in the sixth hour of the day, in the middle of the day, the one who created him, died for Adam / Adem. And returned to his Father, and to our Father, to complete the plan of salvation.

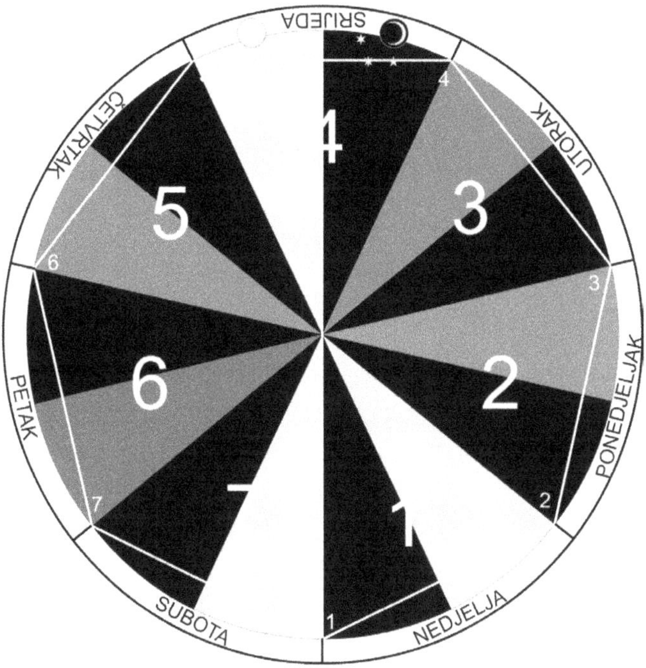

Picture 27 (LIFE AND DEATH THROUGH TIME AND SPACE).

NEDELJA – SANDAY; PONEDELJAK – MONDAY; UTORAK – THUSDAY; SRIJEDA – WEDNESDAY; CETVRTAK – THURSDAY; PETAK –FRIDAY; SUBOTA - SATURDAY

Shows time in days, hours and minutes. Confirms the signals given by God because they are exactly the information for this time, in which we live.

And the time we live in is the time of grace.

When the time has come to an end, that means that there is no more time. We live in a time of the grace of God. The time of grace God gave first to Adam / Adem, and in the grace of God live Adam's / Adem's children today to.

God's grace postponed his death and found a substitute for his sin and gave himself so that man could live forever. Everyone who understands, the plan of salvation, who keeps the 10 commandments of God, who believes that he who is so described in the word of God, who fulfilled and confirmed everything written earlier, that it contains the fullness of everything they will have eternal life.

It's the time of grace that Adam / Adem got.

The second generation received its time of grace before the flood. For 120 years, Noah made a ark, and warned the people that they would be flooded. To return from their sinful ways, to change the way of life in which their idols and their desires were Gods.

They used their time for their downfall. The penalty for sin is death. Any unrepented sin is also death. People before the flood they did not use their time of grace they received, a time of 120 years, while Noah / Nuhh made a ark. Every hammer blow was at the same time a warning to them. But they neither wished to hear about the warning nor understood or took advantage of it and they received their wages which was eternal death.

The third generation had its time of grace. God had mercy on his people in Egypt, and would be sorry for them and brought them out of slavery.

The fourth generation had its time of grace. Because of their disobedience to the law of God and because of their idols they went into slavery to the Persian emperor and were enslaved for 70 years. The LORD, had mercy on them and brought them out by his grace.

The fifth generation had Jesus / Messiah, they did not recognize him because those who lead them, lead them into ruin.
The sixth generation has its times of grace, it is already ticking, it came. It is not known the day or time of Christ's second, coming. But by the events you will know.
A time of 6.000 years, in which people changed hours, days, years and calendars. So that God's calendar cannot be easily or simply revealed because it contains the message and truth.
The, LORD GOD, left and placed signs in heaven to which people and evil angels cannot come, nor can they influence their specific task given to them by the creator.
By them the time is counted, by SUN the year, and the MOON; the days, weeks and months.
God's calendar says we are at the end of (all) time.
6.000 years is near the end. But the day and time of the second of Messiah/ Christ arrival is not known, it is the mystery of God. Even Jesus / Messiah himself does not know. Only Heavenly Father.
After the death of Jesus / Messiah who exhaled on Friday, on the sixth day in the week, there was darkness from the sixth to the ninth hour.
His disciples came and removed him from the cross, at Golgotha, where those sentenced to death died in the most severe torments. Both the Romans and the Jews were surprised to see how quickly Jesus exhaled.
His disciples buried him in a new tomb, in which no one laid, on Friday, the evening of the great Jewish holiday.

In the Gospel of Luke:

Chapter 23:

Verse 44. “And it was about the sixth hour, and there was a darkness over all the earth until the ninth hour.”

And the Sun was darkened, and the veil of the temple was cut in half.

And when Jesus / Isaa, had cried with a loud voice, “Father, into your hands I commend my spirit “ and having said this he exhaled. And when the captain saw what happened he began to praise God, saying, “This man was truly a righteous man.” And all the people who were gathering to watch, when they saw what happened, returned beating into their chest. And all his acquaintances stood afar off and women who followed him from Galilee watched this.

And behold, a man named Joseph / Yusuf a counsellor a good and just man.

As he approached Pilate he asked for the body of Jesus / Isaa.

And he took him off and wrapped him in cloth. They put him in a grave where no one was laid. The day was Friday, and the women who came with Jesus / Isaa, from Galilee, followed Joseph / Yusuf and saw the grave.

And on the first day; Sunday, they came to the grave very early and they brought the spices which they had prepared, and some other women with them. But they found a stone detached from the grave. And entering they did not find the body of Jesus / Isaa. And they were astonished, two men stood before them in shining garments. And when they were afraid, they bowed their faces to the ground, and they said unto them,

'Why do you seek the living among the dead?' He is not here. But rise up and remember how he told you when he was still in Galilee that the Son of Man should be delivered into the hands of sinful men and crucified on the third day.

And they remembered his words.

Returning from the grave, they reported all this to the eleven and all the others.

Verse 12. "Then arose Peter, and ran unto the sepulchre; and stooping down, he beheld the linen clothes laid by themselves, and departed, wondering in himself at that which was come to pass. And, behold, two of them went that same day to a village called Emmaus, which was from Jerusalem *about* six runners ."

And they talked to each other about all these events.

And as they talked and asked one another, Jesus / Isaa approached and went with them. And their eyes were kept from knowing him.

He said to them, " What kind of conversation are you having between each other, why are you unhappy?."

And one, named Cleofa, answered, "Are you the only one of the churchmen in Jerusalem that you have not heard what happened these days?."

And he asked them what was the matter. And they said of Jesus / Isaa, the Nazarene, who he was a prophet, mighty in work and in the word before God and before all people. And how the chief priests and our rulers delivered him to be condemned to death, and have crucified him. But we trusted that it had been he which should have redeemed Israel: and beside all this, today is the third

day since these things were done. And certain women also of our company made us astonished, which were early at the sepulchre; and when they found not his body, they came, saying, that they had also seen a vision of angels, which said that he was alive.
And certain of them which were with us went to the sepulchre, and found it even so as the women had said: but him they saw not.
Then he said unto them, O fools, and slow of heart to believe all that the prophets have spoken: Ought not Christ to have suffered these things, and to enter into his glory?
And beginning from Moses and all the prophets, he told them about all the scriptures the things concerning himself.
And they came nearer to the village to which they were going, and it seemed like he wanted to continue going. And they stopped him, saying, "Stay with us, because the day is almost over and it's close to night." And he went in with them that night.
And when he sat down with them at the table, he took bread and blessed it, and broke it and gave it to them.
Then their eyes were opened and they knew him. And he disappeared.
And they said one to another, did not our heart burn within us, while he talked with us along the way , and while he opened to us the scriptures?
And they rose up the same hour, and returned to Jerusalem, and found the eleven gathered together, and them that were with them saying, The Lord is risen, and appeared to Simon. And they told what things were done on the way, and how they recognised him when he broke the bread.

Verse 36. "And when they were saying these things, Jesus himself stood among them and he said to them, "Peace be upon you." But they were terrified and got afraid, and thought that they had seen a spirit.

And he said to them, "What are you afraid of? And why do such thoughts enter your hearts? See my hands and my feet, that is I myself, touch me and feel me and see, the spirit has no body and no bones, as you can see I have."

And saying this, he showed them his hands and feet. And while they didn't believe it yet, out of joy, and they wondered, he said unto them, "Is there anything here to eat?. And they gave him a piece of baked fish and an honey . And he took it, and did eat before them.

Verse 44. "These are the words which I spake unto you, while I was yet with you, that all things must be fulfilled, which were written in the law of Moses, and in the prophets, and in the psalms, concerning me."

Then opened he their understanding, that they might understand the scriptures.

Verse 46." And said unto them, Thus it is written, and thus it behoved Christ to suffer, and to rise from the dead the third day. And to preach repentance in his name and forgiveness of sins among all nations, beginning from Jerusalem.

And you are witnesses to this.

Behold, I will send the promise of my father upon you, and you sit in the city, Jerusalem until you be endued with power from on high.

And he led them out to Bethany, and, lifting up his hands, blessed them.

And when he had blessed them, he departed from them, and ascended into heaven.
"And they worshiped him and returned to Jerusalem with great joy."
The Holy Gospel of John: Chapter 1:
Verse 1. "In the beginning the Word rages, and the Word rages at God, and God rages the Word."
The Word in the beginning rages at God.
All things were made by him; and without him was not anything made that was made.
There was life in her and the life was the light to people.
And the light shines in the darkness, ; and the darkness comprehended it not.
There was a man sent from God, whose name was John. He came for a witness, to bear witness of the Light, that all men through him might believe.
He was not that Light, but was sent to bear witness of that Light
It was the true one, the one who shines on every human who comes into the world.
He was in the world, and the world was made by him, and the world knew him not
He came unto his own, and his own received him not.
And he that received him gave them authority to be the sons of God who believed in his name.
Which were born, not of blood, nor of the will of the flesh, nor of the will of man, but of God.
And the Word became flesh and dwelt in us full of grace and truth. And we saw his glory as the only begotten of the Father.

John bare witness of him, and cried, saying, This was he of whom I spoke about, he that cometh after me is preferred before me: for he was before me.

“And of his fullness have we all received grace.”

For the law was given through Moses, but grace and truth came by Jesus Christ.

So far we have walked the line of light and the line of time that fit and are made together in the same place for 4.000 years, right in the year, day, hour and time, “Alpha and Omega.” There the firstborn in heaven gave his life for the firstborn on this Earth. To save to everyone who believes in him, to give him eternal life. That the LORD of life and death, rewards by the deeds of all created and all born to the sons of Adam’s / Adem’s and daughters.

And who will come at the end of all time to awaken from mortal sleep one to eternal life and the other to eternal death.

Everything we see around us gives us the same information available to every intelligent being.

The universe testifies to the accuracy of this information.

The line of imperial blood, which ran from Adam / Adem, through the first born sons whose names are written down for accurate information, and are recorded both when they were born and when they passed away, ever since the first man Adam / Adem, Noah / Nuha: Abrahama / Ibrahima: Izaac / Isaac> Jacob / Jakub – Israel: Moses / Muse>: David’s / David’s until Jesus /Isaa: who was born exactly on the year 4.000.

At 1.440.000 days:

At 1.440.000 hours:

At 240.000: 4000 = 60minutes.

The line of time, the line of light, the line of pure imperial blood across Gods chosen people, until a certain date; year, day, and time at which the creator, the redeemer, and the saviour was born. For which all the prophets wrote, for which space and time are witnesses, to which nature testifies, all confirm to us, that all this information has in itself the Alpha and Omega of the perfect movement of all creation from the beginning of all beginnings. And these are the light, the seconds, minutes, hours, days, months, years, as well as thousands of years.

They had direction of light that goes from the first second and emergence of light and life on this Earth. To show us and point us to the most important information that went along the line of light, the line of time, and the line of blood, from Adam / Adem to Noah/ Nuha. And that was also the first period in which the first 1.000 years passed.

Then from Nuh / Noah to the Abraham / Ibrahim; 1.000, years have also passed.

Then from Abraham / Ibrahim to the David / Davud; 1.000 years have also passed.

Then from David / Davud, to Jesus /Messiah; 1.000 years have also passed.

This line and this direction, God has maintained for 4.000 years, everything until the plan of salvation is fulfilled. And the plan of salvation God made for people on the day Adam / Adem arbitrarily violated God's commandment and therefore lost the holiness he had until then.

“Although he was warned, God said to him.“ If you taste the fruit from this tree, you will die that day.”

Because the Creator gave man free will, and Adam / Adem was warned that was the only condition God had given him so that he lives forever in the Garden of Eden. Yet he decided to give his love to his wife and not to the God who created them both.

It was his free will.

1. **THE BOOK OF GENESIS:** Chapter 2:

Verse 15. “And the Lord God took the man, and put him into the garden of Eden to do it and take care of it.”

Verse 16. “And the Lord God commanded the man, saying, Of every tree of the garden you may freely eat.”

Verse 17. “ But of the tree of the knowledge of good and evil, thou shalt not eat of it: for in the day that thou eatest thereof thou shalt surely die.”

CHAPTER 3:

Verse 1. “But the serpent was more subtil than any beast of the filed which the LORD God had made. And he said unto the woman, Yea, hath God said, Ye shall not eat of every tree of the garden?”

Verse 2. “And the woman said unto the serpent, ‘We may eat of the fruit of trees of the garden.”

Verse 3. “But of the fruit of the tree which is in the midst of the garden, God hath said, Ye shall not eat of it, neither shall ye touch it, lest ye die.”

Verse 4. “And the serpent said unto the woman, Ye shall not surely die.”

Verse 5. “For God doth know that in the day ye eat thereof, then your eyes shall be opened, and ye shall be as gods, knowing good and evil.”

Verse 6. “And when the woman saw that the tree was good for food, and that it was pleasant to the eyes, and a tree to be desired to make one wise, she took of the fruit thereof, and did eat, and gave also unto her husband with her; and he did eat.”

Verse 7. “And the eyes of them both were opened, and they knew that they were naked; and they sewed fig leaves together, and made themselves aprons.”

Verse 8. “And they heard the voice of the Lord God walking in the garden in the cool of the day: and Adam and his wife hid themselves from the presence of the Lord God amongst the trees of the garden.”

Verse 9. “And the Lord God called unto Adam / Adem, and said unto him, ‘Where are thou?”.

Verse 10.“And he said, I heard thy voice, in the garden and I was afraid, because I was naked; and I hid myself.”

Verse 11. “God said, ‘Who tells you that you are naked? Hast thou eaten of the tree, whereof I commanded thou shouldest not eat?.”

Verse 12. “And, Adam / Adem said, the woman whom thou you gavest to be with me, she gave me of the tree, and I did eat.”

Verse 13. “And the Lord God said unto the woman, What is this that thou hast done? And the woman said, The serpent beguiled me, and I did eat.”

Verse 14. “And the Lord God said unto the serpent, Because thou hast done this, thou art cursed above all cattle, and above every beast of the field; upon thy belly shalt thou go, and dust shalt thou eat all the days of thy life."

Verse 15. "And I will put enmity between thee and the woman, and between thy seed and her seed; it shall bruise thy head, and thou shalt bruise his heel."

Verse 16. "Unto the woman he said, I will greatly multiply thy sorrow and thy conception; in sorrow thou shalt bring forth children; and thy desire shall be to thy husband, and he shall rule over the."

Verse 17. "And unto Adam he said, Because thou hast hearkened unto the voice of thy wife, and hast eaten of the tree, of which I commanded thee, saying, Thou shalt not eat of it: cursed is the ground for thy sake; in sorrow shalt thou eat of it all the days of thy life."

Verse 19." In the sweat of thy face shalt thou eat bread, till thou return unto the ground; for out of it wast thou taken: for dust thou art, and unto dust shalt thou return."

And Adam / Adem called his wife's name Eve / Hava, because she was the mother to of all people. Unto Adam/Adem also and to his wife did the Lord God made leather dresses, and clothed them.

Verse 22. "And the Lord God said, Behold, the man is become as one of us, to know good and evil: and now, lest he put forth his hand, and take also of the tree of life, and eat, and live for ever."

Verse 23. " Therefore the Lord God sent him forth from the garden of Eden, to till the ground from whence he was taken."

Verse 24. " So he drove out the man; and he placed at the east of the garden of Eden Cherubim, and a flaming sword which turned every way, to keep the way of the tree of life."

Adam / Adem was warned before he sinned. He knew the consequences it will have disobeying the one who created him. And the moment he showed disobedience to God, when he did what God forbid him, it was at that moment he also got to know what is good and what is evil.

At that moment, he took responsibility for all his deeds that he will do in the future as well as be responsible for them because until he knew about good or evil did not even have sin.

Due to disobeying the law of God, he could no longer use Eden garden as his home.

While living in it he also used the tree of life and it made it possible to live forever.

As sinners cannot use the tree of eternal life Adam / Adem and Eve / Hava, had to leave their home and get out of it. And to start living a different life that until then they didn't even know.

God said to them, "With the sweat of your face you will eat bread until you return from the earth which you were taken; you are dust and you will return to dust."

Adam / Adem, until then, was sinless because he knew no obedience and sin.

The penalty for sin is death and someone had to die for him.

God declares for himself that he is gracious, sad, slow to anger, and that his name is love. And he was sorry for the man he created, to die, and out of his love and his mercy for the sinner Hi made a plan - plan, of salvation for this human race. He created a replacement and delayed the sinner's death.

The right to the tree of life is conditioned by obedience.

The transgression of God's commandment excluded Adam/Adem from the Garden of Eden. And the angels with the flaming sword are set to guard the approaches to the tree of life so that man would not reach for its fruit by tasting it, he would become an immortal sinner, eternal, and thus sin itself.

Taking fruit from that tree was conditioned by keeping all of God's commandments.

By his fall into sin due to disobedience Adam/Adem lost all right not only to the life giving fruit of the tree but also to its leaves which serve to heal the nations.

Obedience through Christ / Messiah gives a man perfect character and right to the tree of life.

The conditions for regaining the right to the tree of life are clearly set out in the book "The Revelation of John".

CHAPTER 22:

Verse 13. "I am the Alpha and the Omega, the beginning and the end, the first and the last."

Blessed are they that do his commandments, that they may have right to the tree of life, and may enter in through the gates into the city.

Verse 16. "I Jesus have sent mine angel to testify unto you these things in the churches. I am the root and the offspring of David, and the bright and morning star."

The book of the prophet Isaiah says that a savour, redeemer, and deliverer will be born from the root of David.

CHAPTER 22:

Verse 22. “And the key of the house of David will I lay upon his shoulder; so he shall open, and none shall shut; and he shall shut, and none shall open.”

Verse 23.“ And I will fasten him as a nail in a sure place; and he shall be for a glorious throne to his father’s house.”

When Abel, the son of Adam/Adem, brought the firstborn of the flock, he thus acknowledged god, not only as the giver of earthly blessings but also as the one who gave us the saviour.

The sacrifice of sacrificial animals was accepted as a temporary substitute because of the fallen and sinful man, because his guilt, could no longer go directly before God with his prayers because his transgression of the divine law raised an impassable barrier between the holy god and the transgressor.

God created a plan of salvation according to which the death sentence should fall on Christ as man’s substitute

The plan of salvation must have provided for the shedding of blood, because death is an inevitable consequence of man’s sin.

(The sacrificial offerings were a foreshadowing of Christ’s sacrifice).

In the slaughtered sacrifice, the man had to watch the temporary fulfilment of the words- death.

Ever since the day Adam / Adem was cast out of the Garden of Eden, he began to offer sacrifices for his sin because Adam / Adem fully understood that only through sacrifice, a slaughtered lamb that is slaughtered because of his sin, which is a signpost, that someone sinless will have to die so that he and his descendants who would be born could live and return to the Garden of Eden again.

Jesus / Isaa stepped on Adam / Adem path. What kind of love, or humiliation, the emperor of glory offered himself to be descend to the fallen human race.
He was ready to step with his own foot on Adam's / Adem's path to take the nature of a fallen man and enter into battle with a strong enemy who triumphed over Adam/Adem in triumph.
To defeat Satan and thus open the way for all who believe in him to be redeemed from Adam's shameful failure and fall.
The execution of the pronounced sentence has been suspended. God refrained for some time from the definitive execution of the death penalty imposed on man. Satan made it clear to himself that he will break the connection between heaven and earth. But he would be greatly mistaken and very disappointed. The Father himself surrendered the world into the hands of his son to redeem it from the curse caused by Adam's shameful failure and fall into sin.
The replacement occurred at the same time as man yielded to Satan's temptation by doing what God said he should not do. Christ/Messiah the Son of God placed himself between the living and the dead, saying: "Let the punishment fall on me, I put myself in man's place to give him another chance". Every man was given another chance through a long grace and through a holy sacrifice offered on Golgotha through faith.
Keeping the law of God, the (sheriat), the ten commandments of God, as a sign of respect to God and his law, in order to return again to Garden of Eden.

Through the following images that confirm all this information sent by the universe since the creation of the world and life on this earth are aimed at confirming these directions and instructions sent to us by God through all creation

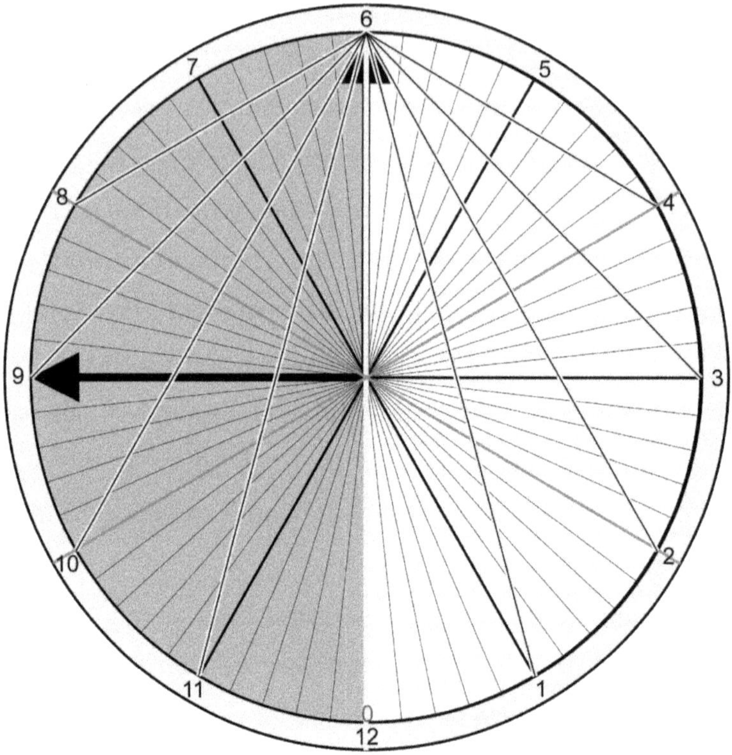

Picture 28 (LIGHT AND TRUTH IN ALPHA AND OMEGA).

It shows a line of light passing through the darkness of the first second of the origin of life on earth.

Here it is seen exactly 12 hours from the first day of the creation of light the truth illuminates the Earth and through time has shown the way, truth and life.

And on the hour, of time from the first week at the place where Friday the day on which God created Man, and on which he also ended up making life on this Earth. From here, the line, light or the truth line.

Picture of light and first second:

In this picture we see light that separates darkness from light from the first second because light is the unit by which time is counted and which reveals darkness. Out of the darkness a light appears and shows the truth and the day that each individual sees the path he is taking or should take.

From the moment that light was created the light will not disappear , neither will the time that counts our days as well as the days of everything created on this earth. Everything created has its time and the length of its time. Everything we see and do not see passes and is renewed in perfect order of doing and dying, thus creating the perfect dimension and the shape of Alpha and Omega.

This picture shows a 360 ° circle and in that circle it is important to notice that we find ourselves like everything on this earth.

This circle of Alpha and Omega is also the limit of our existence and our limits of free movement. Everything beyond these boundaries God created for specific reasons, for the life of this earth and for the perfect functioning of our entire galaxy.

But all these instructions we will see in the following pictures as well all written information, confirming the truth and show the direction of the path, truth and life.

Here the plan of salvation for sin that Adam / Adem fulfilled by his disrespect of God's law, (sheriat), and received punishment

for sin. And the penalty for sin is death.
That's where the creator of man came and gave his life for man so no one who believes him perishes but has eternal life.
This picture also tells us that time as a second formed in the lower part of the circle of Alpha and Omega. First there was zero then from zero the beginning passes light through the middle of space as well as time and reaches 6 o'clock in the middle of the day. And the day counts 12 hours.
This means that every night has 12 hours, as every day has 12 hours. This information says that noon occurs at 6th. six o'clock.
Wednesday (middle) of the week is Day 4th. of creation. It's in that place that God on Wednesday, in the middle of the week, created the Sun, and the Moon and all the Host of the heaven.
On that sixth hour the Sun replaced the light of God, which had shone, and in which was created : Light, time, water, earth, truth and nature. In that place the light of God showed itself to the world, because God is light. And in the light is life and truth.
All images that the Universe sends and that serve us for information show Jesus / Messiah on the cross of Golgotha.
His head is CENTERED in the middle right at the zenith in the middle of the day and in the middle of all time and in the middle of all space.
His body halved both time and space through the middle.
With his right hand he touches the eighth hour and with his left hand he touches the fourth hour and connects God's light with God's creation the sun and binds the 4th commandment of God with 4 days a week.

It connects 4.000 thousand years with the fulfillment of the accuracy the promises he made to Adam / Adem, that the sinless would come and be sacrificed for his sin. For 4.000 years, the sacrifice for sin was offered through God-chosen people.
From, Adam / Adem and his descendants to Noah / Nuh who By God's instructions, they lived in the sinful world of that time. God passed him over the flood to be the father of a new generation
On the 10th generation from Noah / Nuh, Abraham / Ibrahim was born; whom God called and who willingly responded to God's call and set out from Mesopotamia to the land of Canaan with his wife Sarah and his father Tara and his uncle son, Lot, and with everything , what he had; believing God and respecting him. He came to Canaan in the place of Luza , and God changed the name of Luza to Bethel. Abraham / Ibrahim believed God and it was written in his justice.
From Abraham / Ibrahim, through his son Isaac, grandson Jacob/ Jakub, his descendants to the Emperor David / Davud through God-elect, faithful, and God obedient people to the birth of Jesus / Messiah.
On Friday, when man was created, as the crown of God's creation, on the sixth day of the week, at the same place at 6 six o'clock on the sixth day of the week on Friday, after 4.000 years he gave his holy life.
Creator of life.
Creator of light.
Creator of time.
Creator of all that is seen and not seen, with his love, grace which

he had all time. He gave himself up for the fallen human race. And in his long grace, which he set in heaven, he promised and confirmed Noah / Nuhu not to be afraid of the flood and that there will never be a flood again to flood this planet.

He also shows to all of us in the long grace that he has suffered our sins for a long time, and that he is slow to anger but also that the punishment for sin is death. We are sinners from birth, through our father Adam / Adem and that with our merits we cannot return back to the Garden of Eden. That sin could only be bridged by the one who created us. Only he has the power to defeat the sinner, and bring back the fallen through his own sacrifice and his own blood instead of the sacrificial lamb that 4.000 years pointed to that event.

This event is a connection between God and Man, for Christ who is born of God, in the spirit of God, and thus again confirms that He is not just a prophet like other prophets.

Because, all the other messengers lie in the Earth , and await the day of judgment. Everyone will stand Infront of God and be accountable for his deeds for they are all born of man, it means that they have inherited sin from birth, but Christ has no father from the Earth, His father is in heaven and gave him all the power to destroy the death that a man deserves, and destroy the lie.

The father of lies, the human executor is Satan, the devil who stands opposite to God and with his lies he tries to destroy God and Man.

God created the light in which he revealed the lie, showed the truth through everything created in Heaven and on Earth everything shows the creator, the Redeemer and Savior.

All nature, time, light, all numbers and all letters say that God is one and only and that there is no other, and that he alone created all. There is also the seal of God on all creation.
God who said to Moses / Musa: “I am the LORD God, and have no other besides me.”

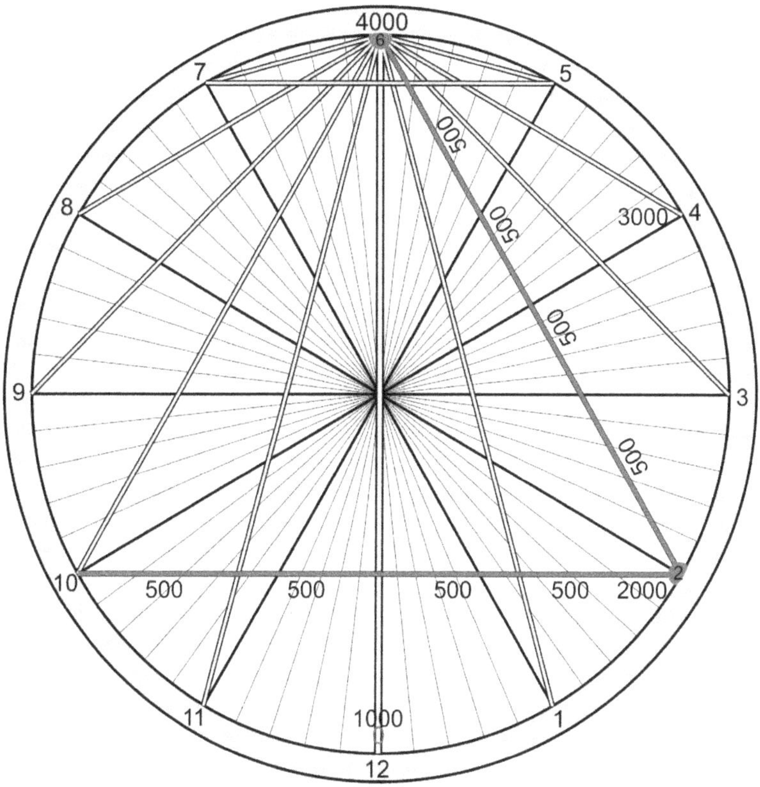

Picture 29 (LIGHT AND TRUTH THROUGH THE BLOOD LINE).
We see light from one day. Having seen from the past information that the day is 12 hours and those hours are counted as days of the week. From zero, from right to left, and that zero and number 12 are located at the bottom of a 360 ° circle.

The middle of the day is at the sixth hour of the day. This picture shows one day, the Sun at the zenith from which light illuminates the earth and touches its own heat of life all created on earth.

This picture shows the sixth hour in the center connected to all other hours of the day also confirms that light from zero in seconds passed through all twelve hours and that every day has twelve hours of light and twelve hours of darkness. The time that divides light into night and day bears the stamp that divided truth from lies, death from life and that exactly half of all the time this planet had was half-darkness and half light. In the middle of that half, the creator, savior and redeemer of the human race was sacrificed, and he was born in human form so that the sacrifice would have complete proof. That proof is another seal of the truth of the information we have from the universe

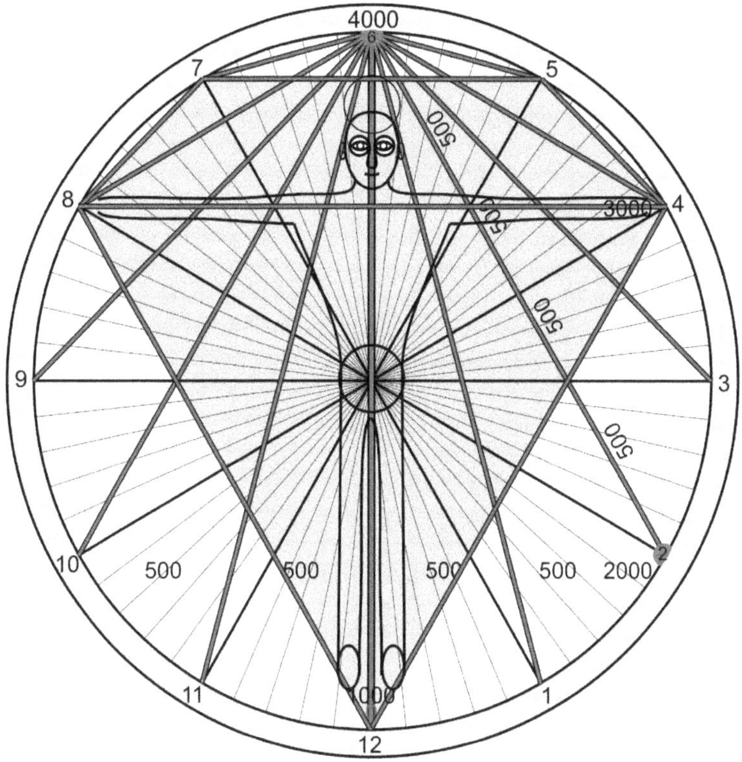

Picture 30 (THE BLOOD OF CHRIST'S IN TIME I SPACE OF 360 DEGREES).

In this picture:

We have light that has engulfed our planet, then hours, days, months, years and thousands of years. Then the space is seen as a 360 degree circle.

You can see the exact time since the creation of full life on earth, all six working days as well as the day and hour when man was created but it is shown as 500 years instead of 5 minutes to make it easier to understand; time and space.

Below you can see the 4 x 500 years leading from Adam / Adem to Abraham / Ibrahim.

That's 2.000 years.

Then from Abraham / Ibrahim to Jesus / Messiah 4 x 500 years = 2.000 years.

It's all together: 8 x 500 = 4.000 years, ending at six o'clock, at noon, at the zenith, the center of all environments

From this centre comes the light of all time and illuminates the truth in which the Jaspis or Alem- Stone is found on which the whole Universe rests and through which everything that goes through that exists passed. It is a stone, an angle that the MASONS of truth have rejected and without which the truth cannot stand. He reveals himself in space and time, and the Sun and light witness and show his glory and worth.

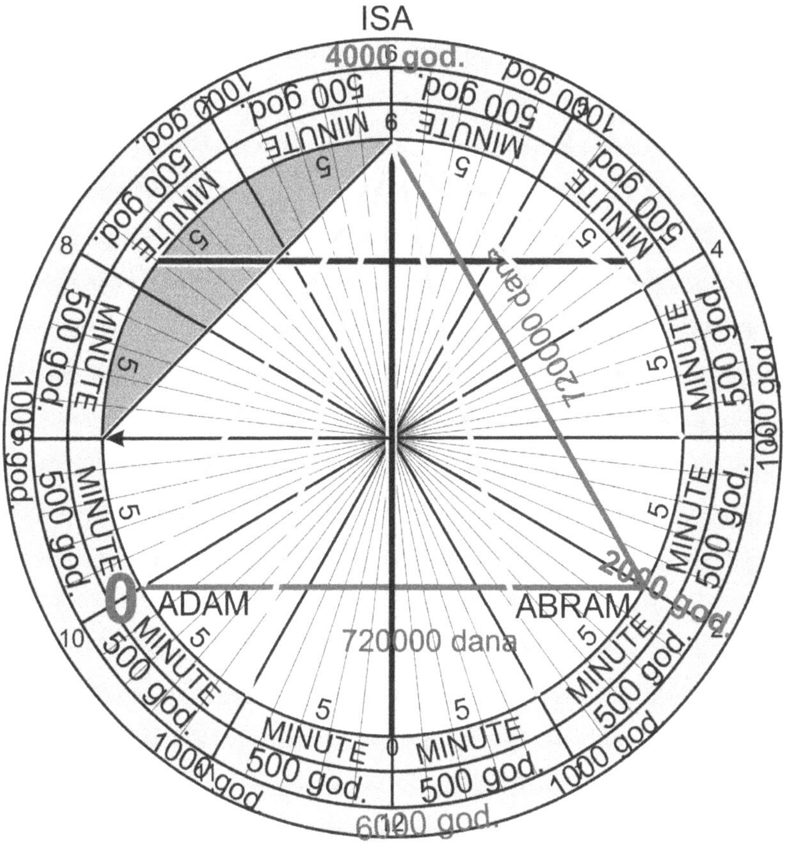

Picture 31 (BLOOD LINE THROUGH THE OLD ERA).

This image reveals a line of blood that was kept from mixing with other blood because Jesus/ Messiah was to be born of Adam's / Adem's imperial blood. That is why God has guided the exact line of blood since Adam / Adem over these 4.000 years.

They have information and the names of the people who passed on that imperial pure blood to the one who was to be born of it. Adam/Adem was the emperor of this planet until he had to hand over his royal crown for sin to the one who deceived him because of his disobedience and became a slave to sin.

But his imperial blood remained pure and was transmitted all the way to a certain person, who was born of the Holy Spirit through a certain person of the royal family.

This origin was prophesied 600 years before the birth of Messiah's anointed over the prophets.

The Saviour will be born in Bethlehem of the Holy Spirit and he will return to his father, who created him.

This picture you can see the blood line leading to a specific day and time , that is again a confirmation of its importance and holiness.

Jesus / Isaa was born exactly 1.440.000 days, from the origin of life on earth.

Adam / Adema was created by God at the 144th. hour, from the creation of life on this Earth.

He was created on Friday, the 6th. sixth day, from the creation of life on this Earth.

Jesus / Isaa died, on the Friday, on the 6th. (sixth) – (hour) a day, right in the middle of the day: are fulfilled of creating life on this Earth.

The number of days that are fulfilled and to confirm that Jesus came exactly for the Man he created in the 144th hour of creation to replace his death by own death and to bring him back to life with his own life.

Gave another chance to all the descendants of Adam to wash away their sin through his holy blood and through the love of God.

By keeping the law of God, the 10 commandments of God, which he gave to Moses / Musi on Mount Sinai; gave the opportunity that to be able to come again to the Garden of Eden and have eternal life.
As Christ was dying from the (6th) sixth to the (9th)) ninth hour, the darkness would be on the Earth and the Sun shielded its light from this scary event. In those hours when the plan of salvation for everyone was fulfilled and any individual who seeks God and who like all holy people, wants to please God, they have a chance through holy blood and that holy sacrifice.
In those moments, the rain began to fall and the blood of Christ was in the storm, carried to the rivers and rivers to the sea. So the blood of Christ covered the whole Earth.
It encompasses all nations, tribes and tongues. God's grace descended on each individual because God would not be pleased that the man he created would disappear from the Earth.
And created signs in heaven and Earth to show time and that in time we can prove and show the accuracy of information. As if time has its time and its limitation and that it is also transient.
That time is limited by time and that time will come to an end. Everything created is limited by time. This means that everything will pass as God created everything in the first week of creation. And that everything is repeated in 6.000 years through its Alpha and Omega.
Everything is born and dies and then again. Everything has its appointed time.

Joseph / Yusuf was like a faithful illustration of Jesus. And he came to his own, and his own received him not. He was also rejected and despised because his right actions and his consistent life full of renunciation, he was a constant rebuke to those who presented themselves as pious and whose lives were perverted.
Joseph's / Yusuf's honesty and moral chastity were fiercely attacked. When the woman tried to lead him into sin and failed she got even angrier. She falsely testified against him. He suffered while innocent. For his righteousness and moral integrity he was most cruelly thrown into prison.
His brothers sold him as a commodity to his enemies for a small sum of money. Likewise, one of his disciples sold the Son of God to his bitter enemies.
Jesus remained gentle, calm and holy.
He lived a life of incomparable renunciation of goodness and holiness. No guilt and no evil could be attributed to him, yet false witnesses were bribed to testify against him. He was hated for faithfully rebuking and condemning sin and depravity.
Joseph / Yusuf brothers took off his colorful robe before throwing him into the dagger, just as they threw a dice for Christ' clothes to whom it will belong.
God allows trials and temptations to come. These are the means by which those who claim to be faithful to God are tested and tempted. We read in the Word of God that puts all his elect to the test. This means that he allowed trials to come that would test their faith and lead them to expect help from him. If they

approach him in their temptation in faith and prayer, he gives them the strength to overcome all temptations.

We read in the Word of God that God put Abraham / Ibrahim to the test and his test was the hardest.

Had he not maintained then he could never have been registered as the father of the faithful. If he had deviated from God's explicit commandments, the world would not have had such an example of indisputable faith and obedience. The message provided in this way shines through the ages so that we too can learn. Nothing is so precious that we could not give to God.

God intended the offering of Isaac to be a foreshadowing of his son's sacrifice

Isaac was a forerunner of the Son of God who was truly sacrificed for the sins of the world. God wanted (Abraham-Ibrahim) divine message of salvation to be revealed to man and imprinted deep in (the soul) to turn that truth into reality and to test his faith; he asked him to slaughter his beloved only child and offer him for sacrifice. The purpose of all the mental sufferings that (Abram-Ibrahim) had to endure during that dark and passionate ordeal was to imprint in his soul as deeply as possible the idea of the plan of redemption of the human race and to make it easier to understand.

The Creator gave us all this information to make it easier for us to realize that the price God had given us, was imprinted in the depths of our souls the plan of salvation for each individual.

The love of God created the plan of salvation so that the Son of God, the first created in heaven, would be born on Earth and justify the first created on Earth with his blood, Adam/ Adem, and through his sacrifice return him to eternal life in the Garden of Eden. And all his descendants who through this sacrifice ask the creator to justify them.

"10 Commandments of God given on Mount Sinai to Moses / Musa in hand"

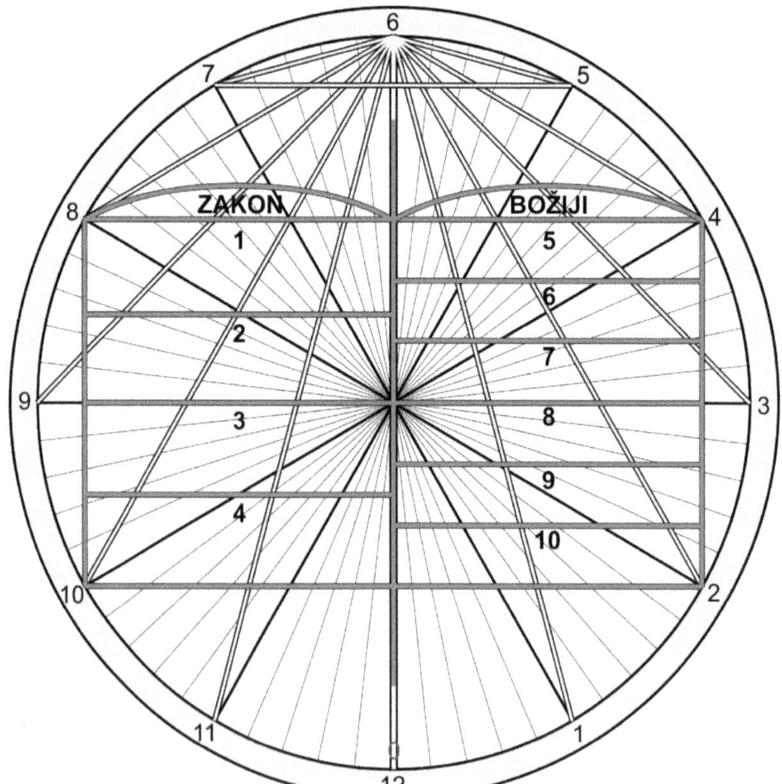

Picture 32 (THE LAW OF GOD IN THE MIDDLE OF TIME AND SPACE).

Shows time of one day 12 hours, time of one year 12 months, and time of 6.000 years. We see the light from which The Creator creates all the time and encompasses all Earth from 1. second up to 6.000 years.

You can see the middle of the day when God, in the middle on Wednesday, created the Sun and all the host of Heaven, the rays that shine upon the whole earth, the truth, and life eternal.

This picture clearly shows two stone slabs, the law of God, which The Lord God wrote with his finger and gave Moses / Musa to put them into the ark of the covenant.

May God's law be our refuge and our defence in obedience to the law, through faith in the sacrifice of the lamb slaughtered. Since the beginning of the world: Jesus-Christ the Messiah.

It can be seen here that at exactly 4.000 years, on the exact day, hour and minute through the middle of the planet, in the direct light of God at mid-day, mid-week, mid-month, and the middle of the year, the law of God stands in the midst of 6.000 thousand years.

He stands and points to the cross, on Golgotha, and is connected with death with the Saviour and with eternal life, and that they are in the same place because they are connected by the same line through which the direction leads to eternal life. Obedience to the law of God, which the first man refused (by) faith in purification through the sacrifice and blood of the lamb, together have the power to cleanse us from every flaw and shortcoming in our character through knowledge of truth and faith.

From this information we see that the law of God has been in heaven before. That it is there now and that it will always be there. Thus he says that the law of God is eternal and that no letter from it will ever be changed or removed. Jesus did not come to change the law but Hi,came to fulfill it.

We see that the law of God is eternal and that there is a law of obedience to the Creator in heaven. So has Jesus been in heaven since eternal times. He came to earth to defeat sin and the originator of sin and disobedience, and to fulfill the law of God with his obedience, and to restore man to the law in faith in God's love and grace, bringing him back to eternal life.

All 10 words of God have the same power. They all ask people the same understanding, that we have love for God and love for Neighbours as for ourselves. Only through such love can we get rid of our flaws that separate us from the love of God, which we can only gain through obedience. This picture shows the great light with which the law of God is illuminated or the truth that illuminates it so that everyone can see, read and understand it.

The law of God has 10 commandments found on two stone tablets. On the first stone slab are the 4 commandments of God and on the second the 6 commandments of God.

The first four refer to God and describe him as the creator of everything visible and invisible, and show a reference to information that says who is the creator of everything visible and invisible around us and who maintains all life on this planet. For all creation is connected by joint work.

Through this joint work, where everything created complements in its function and which God has ordained for every creation. And every creation is made for its own function. Through which the life cycle is maintained and created.

Every creation has its exact function and nothing created by God is not created without a specific reason and function. Everything created has its Alpha and Omega, the beginning of his function, which is renewed by the will and wisdom of God to the point of certain time. In the beginning of making life on this planet God created the conditions for man to live on a specific time and space that is special Alpha and Omega in his creation and with the task for which he was created.

Everything was written by the Creator, who is described in these first four commandments with his finger on the stone slabs that have the task to

eternally exist and which show that the law of God is eternal as God is also eternal.

The second stone block contains the six commandments of God which refer directly to people and refer to right behaviour to each other. In obedience to that law, people will bring themselves peace, health and a long life. Then through obedience he will show respect to God and his word.

Thus he will receive a blessing on this earth as well as the blessing God gives to those who keep his law and its provisions. It will give them power to eat from the tree of life that is in the midst of the paradise of God.

These six commandments are used on Earth today and are the centre of all state laws, but these first four commandments have lost their significance because people want to believe in God according to their own measure and their own desire.

However, God is one and does not change according to our desires.

We need to turn to his wishes and his law. In that is the whole truth. The law of God, given to Moses on Mount Sinai and it is given for all nations, tribes and languages.

"The original law of God is in the ark of heaven."

(Sister White) in the Bible commentary, on page 69 says: "I warn you, do not oppose your influence to the commandments of God."

This law was written by God himself in his heavenly temple. People they can tread a copy of it here on Earth but the original is kept in the ark of God in heaven. Above that ark , above the law there is the throne of grace or the throne of God. Before that the ark stands Christ interceding for man.

"The law of God is kept in the ark, in heaven."

And Christ, saying this to Moses / Musa on Mount Sinai, gave him two testimony plates, 2 stone tablets written with the finger of God. Nothing from what is written, on these plates, cannot be erased. The precious record of the law was put in (the ark of the covenant) and still is located there. Surely hidden from the human race.

But God will at the time appointed by him, take out these stone tablets to be a testimony to the whole world against, general neglect of his commandments and against the idolatrous observance of false Sabbaths.

There is abundant evidence of the immutability of God's law. It is written with the finger of God and will never be erased or annulled. The stone tablets with the law were hidden by God in order to bring them out on the great day of judgment exactly as he wrote them.

"And it says in the word of God, The court sits and the books are opened."

And when everyone receives a salary as it is written in the books according to their deeds, then the stone tablets that God will keep hidden until that day will be shown to the world as an infallible measure of justice. Then men and women will realise that keeping God's perfect law was the greatest precondition for their salvation. No excuse will be found for sin according to the righteous of that law. People will receive judgment; eternal life or eternal death.

Christ is the glory and holiness of the law, the glory from which the face of Moses / Musa shone; It was a reflection of justice in the law. The glory of the law lies in its incarnation, in Christ.

The law, by itself, has no power to save man. He himself is, without splendour unless; Christ is represented in him as the fullness of justice and truth.

The Ten Commandments of God is the most important document for all who live on Earth.

In the middle of the law is the fourth commandment and in it is the name of the creator of Heaven and Earth. As well as the knowledge that the seventh day is Saturday.

It is a sign or seal of God as well as an explicit explanation of God that on that day on the Sabbath there is a special respect for God and his law.
The Sabbath is the day when God, after all that he had created, had left the day to watch and enjoy all that is created.
He was pleased with all that he had created and consecrated that day and blessed it and said, "It is a sign that people will recognise me as creator, it is a sign that divides those who give thanks and who give recognition from those who do not notice at all."
The Gods of wood and stone did not create anything, everything was created by God who showed and proved that he is the creator of everything and that life and death are in his hands. And that there is no other God but him, and that everything is limited by time and that time passes very quickly. Let us not attach ourselves to fleeting things or to the fleeting riches that moths and time eat, but to attach ourselves to the truth of God that leads us to where we should always have been in the Garden of Eden, our lost homeland.
God has shown in his truth that anyone who wants to know the truth can in a very simple way find it and understand it.
No one will have an excuse for his eternal ruin because in his life in his time he had one and only one task to find the way to the truth and the life that leads to eternal time. Through all the signposts we have around us, and everything we see around us, it says the same thing, God did not create us for no particular reason, as if we were not created for this short life on this earth. God created us to live forever even though our father Adam/

Adem lost eternal life and we die we all have another chance from God to seek forgiveness of our sins that though his grace and sacrifice on Calvary, through God's love and Christ's death, we ask for the forgiveness of our sins, inherited from the fathers as from forefather Adam/Adem .

And through the blood that is shed for our sins our sins are washed and cleansed through the recognition of that sacrifice and by keeping the law of God and to take advantage of that second chance of grace and be one of those who will be returned back to eternal life again.

This promise was made to us by the Creator himself, the Redeemer and the Saviour. It is the Creator who created us from the dust of the Earth and gave us our breath of life, and his name is the LORD GOD.

The same holy person called the Redeemer is the one who redeemed us with HIS HOLY LIFE, and protects us from eternal death and by his holy life he paid for our eternal life. That everyone who believes him should not perish but have eternal life.

The same holy person is the Saviour who saves us from eternal death and who will lift us up from the graves when the time is set for that which will weigh us on the scales through its law, through the ten commandments of God, through the faith we had in him, in the belief that he was a sacrifice for our sin, in faith in his love and promise that no believer perishes but has eternal life. To justify us and bring us into the kingdom of eternal life. All this happens through one person who is the centre not only of this Earth but of the whole Universe because everything created shows and says that Christ is the centre of all centres and that he holds and maintains everything created and that he can recreate everything as many times as he wants.

Everything created says he is (The Alpha and Omega) of all creation and that therein is fullness and the fulfillment of everything visible and invisible, as well as what will come.

All the numbers, all the letters, all the nature, all the people, everything has a seal on it from our God and only God who presented himself to Abraham / Ibrahim by the name, (Father Almighty), and to Jacob / Jacub by the name (LORD GOD).

There are still many names in the holy books, but here he showed his name correctly. On some places, he said his name was Emanuel or Michael he then says about himself that his name is love, grace and patience.

Moses / Musi on Mount Sinai, when he received the Law of God and the first coming of Christ is was also shown.

He saw the work of Christ / Messiah, and his mission on Earth. When he had to take on human nature, to act in the world as a teacher and educator, and ultimately to give his life for the redemption of Man.

When the perfect sacrifice for sin was offered, the sacrifices and offerings that symbolically indicated the work of the Messiah were to cease with the first coming of Christ / Messiah. Part of the uncertainty has been removed and the darkened understanding of his people has been flooded with a tide of light.

When Moses saw in a vision the day of Christ and the new living

way of salvation opened by his blood he was overcome with enthusiasm. The glory of God was in his heart.
The splendour of divine glory that accompanied the proclamation of the law of God, reflected so prominently on his face.
Beside the ark with the law of God in heaven there are standing the living Angels.
And the ark with the tablets of God's law in the earthly sanctuary, which was made after the example of the real ark that is in heaven.
And up by the ark of heaven stand the living Angels. Each of them shadows the throne of grace with one of their raised wings and with the other hides his face as a sign of awe and calm.
The curtain in the temple, renewed every year, at the moment of Christ / Messiah's death on Golgotha was torn up exactly at the moment when the priests in the temple performed the ritual service in front of the curtain, which divided the Holy from the Holy of Holies.
The ground shook on him , and the massive lavishly crafted curtain, renewed every year, was torn in two from top to bottom with the same bloodless hand, that wrote those fateful words on the wall of Balthazar's imperial palace.
Because of the sacrifice of Christ / Messiah on the cross of Golgotha, the plan of salvation was fulfilled and the slaughtering sacrifices ceased because when the invisible hand tore the curtain which divided the Holy from the Holy of Holies, at that moment lamb escaped from the priest, under the knife.
And just at that moment Jesus / Isaa died , on the cross of Golgotha, a sacrifice would be offered in him and through him. He has been a slaughtered lamb since the beginning of the world.

All the sacrifices who were slaughtered until that day had the task of opening our eyes and showing that the holy sacrifice will come through his life as well. And through the law of God, keeping It and obeying his every commandment, because in them and through them we respect the Lawgiver God.

That is the only condition and second chance God has given us: both through his grace and through his sacrifice made it possible for us to re-enter the Garden of Eden.

The law of God written with God's finger on two stone tablets and was given into the hands of Moses / Musa, the eternal law by which the human race will be judged on the Day of Judgment. And it reads: (first four) commandments that are on the first board.

---

1).The First Commandment of God:

"I am the Lord your God, who brought you out of Egypt, out of the land of slavery. You shall have no other gods before me."

2).The Second Commandment of God:

"You shall not make for yourself an image in the form of anything in Heaven above or on the Earth beneath or in the waters below. You shall not bow down to them or worship them; for I, the Lord your God, am a jealous God, punishing the children for the sin of the parents to the third and fourth generation of those who hate me, but showing love to a thousand generations of those who love me and keep my commandments."

3).God's Third Commandment:

“You shall not take in vain the name of the LORD God for the Lord will not hold anyone guiltless who takes his name in vain.”

4).The Fourth Commandment of God:

“Remember the Sabbath day, to keep it holy. Six days shall thou labour, and do all thy work: But the seventh day is a Sabbath of the Lord thy God: In it thou shalt not do any work: thou, nor thy son, nor thy daughter, thy manservant, nor thy maidservant, nor any stranger that is within thy gates:

For in six days the Lord made the heavens and the earth, the sea, and all that is in them, but he rested on the seventh day. Therefore the Lord blessed the Sabbath day and made it holy.

These are the four commandments of God that are on the first stone boards and they relate directly to God. Where God showed his name.

This same law also fits to the order of holy beings.

The Sabbath from the fourth commandment was introduced as the weekly day of rest while still in paradise. Since God created the world and at the end created Man.

God introduced the Sabbath as a day of rest for man’s sake.

After Adam’s, sin and fall, nothing of God’s law was taken away. The principles contained in the law of the Ten Commandments of God existed before the fall and corresponded to the order and character of the holy beings. After the fall into sin, these regulations were not changed in principle, but only with some additional rules adapted to the living conditions of Man in his fallen state.

God's law, which has existed since before the creation of life on this Earth, consists of two great principles:

1. Love the LORD your God with all your heart , with all your soul, with all your mind and with all your strength.
2. And the other is love your neighbour as yourself.

There are no other commandments or greater than these. These two principles include the first four commandments, which show Man's duties to God.

The other six principles show his duties to his fellow men.

These principles are more pronounced and worded to Man after his fall, so that they correspond to the circumstances and condition of fallen human beings. This became necessary because the human mind weakened in the midst of transgression.

God's law existed before the creation of man because if there was no law, Adam / Adem, could not become a transgressor. After Adam / Adam transgression, the principles of the law have not changed - its rather than just being clearly sorted and expressed so as to suit, a Man in his fallen state.

Christ, in agreement with the Father, introduced a system of sacrifice. A sacrifice that symbolically pointed to the great and perfect sacrifice of the Son of God. Regulations were given to keep the Decalogue, for the sake of uninterrupted continuation the transgression of moral law in Sinai was restored to a magnificence that required the deepest awe.

Christ / Messiah gave Moses / Musa various religious precepts to be followed in daily life. These ordinances were given explicitly for the purpose of guarding the law of the Ten Commandments. That law does not represent the shadow of the symbols that ceased at the death of Christ, but is obligatory for people in all ages until the end of the world.

Every precept of the law of God is spoken for our own good and our guidance in the direction of eternal life.

God in his word has pronounced so many blessings for those who keep his commandments. Should we then live as God's chosen people, or should we trample on his law, saying that it is no longer required?

To claim that God has abolished the law is the same as to claim that he has destroyed himself.

Each rule of the law bears the mark of the infinite.

It has been made aware of the duty and obligation to obey this law and should be the essence of this last message of grace to the world.

God's law is nothing new, it does not create holiness but makes holiness known and makes it familiar.

It is a set of principles and regulations in which grace, kindness and love are expressed. It shows the character of God's fallen human race, and Man's duties clearly stand out.

The ten commandments of God, ten promises of God, each of them contains one command or prohibition in so far as we obey them as the law that governs the entire universe represents for us ten promises.

God says:

"If you have love for me, keep my commandments."

The essence of God's law is given here in the most concise form, and the conditions of salvation are set out for each son and daughter of Adam / Adem.

The ten holy regulations spoken by Christ / Messiah himself on Mount Sinai represent the revelation of God's character and make it known to the whole world that he has every authority over all human heritage.

This law is of the greatest love that could be shown to man in his ten precepts and is the voice of God speaking from heaven to the human soul in promise:

"Do this and you will not come under the power and control of Satan."

The Lord set up his holy commandments as a defensive wall around his creatures and those who, thanks to that, are saved from sin, everlasting death and can become participants in the divine nature. Their concepts and notions of duty will be clear and pure.

The false notion of God's attributes also represents idolatry .

Do we serve God or Ball, the living God or idols.

Satan places his throne between Heaven and Earth. He caused Man's fall and since then he has been constantly trying to completely erase God's form in Man in order to imprint his image on the Human heart.

Overpowering the guilty, he claims the right over them as over his subjects, and he has absolute imperial power over them. He cannot overthrow God from his throne, but by introducing various systems of idolatry, he places his throne between heaven and Earth between God and Human worshipers.

The loyalty of each individual is tested just as Adam / Adem and Eve / Hava were tested in Eden just as the tree of the knowledge of good and evil stood in the middle of the Garden of Eden so the Sabbath commandment was placed in the midst of the Decalogue.

Of the forbidden tree, God said, "Do not eat and do not touch it, lest you die."

Of the Sabbath, the Lord God says, "Keep the Sabbath because it is holy to you, who would desecrate shall surely be put to death."

Like the tree of knowledge, it was a test of Adam / Adem obedience, so the fourth commandment is a test, by which The LORD today checks the loyalty of each individual.

The Adam / Adem experience should serve as a warning to us at the end of time.

Just as the forbidden tree of the knowledge of good and evil was in the midst of the Garden of Eden and by which Adam's obedience to God's law was tempted so the fourth commandment stands in the middle of God's law and we are tempted by it.

And these are the six commandments of God, which are on another stone tablet, and they apply to our rule over our neighbours. Close are all who live around us from parents, brothers, sisters, family. But all those who live near us are also close. The Lord asks us to have great respect for all our neighbours because only in this way can we live in peace and peace brings health, mood and a long life. Thus we see how the Creator gave us instructions and if we follow them we will have a long life and health on this Earth.

This life is followed by a return to the lost Garden of Eden, Paradise.

The six commandments of God which are on the second stone slab are:

And they read:

5). Honour thy father and thy mother: that thy days, may be long upon the land which the Lord thy God giveth thee.

6).Thou shalt not kill.

7).Thou shalt not commit adultery.

8). Thou shalt not steal.

9).Thou shalt not bear false witness against your neighbour.

10). Thou shalt not covet thy neighbour's house, thou shalt not covet thy neighbour's wife, nor his menservant, nor his maidservant, nor his ox, nor his ass, nor any thing that is thy neighbour's.

And the Lord said to Moses / Musa, "Tell the Israelites the sons of Jacob this: 'You have seen for yourselves that I have spoken to you from heaven: Do not make any gods to be alongside me; do not make for yourselves gods of silver or gods of gold. Make an altar of Earth for me and sacrifice on it your burnt offerings and fellowship offerings, your sheep and goats and your cattle. Wherever I cause my name to be honoured, I will come to you and bless you. If you make an altar of stones for me, do not build it with dressed stones, for you will defile it if you use a tool on it. And do not go up to my altar on steps, or your private parts may be exposed.'

5. BOOK OF DEUTERONOMY

CHAPTER 30:

Verse 19. “I Call Heaven and Earth, to record this day against you, that I have set before you life and death, blessing and cursing; therefore choose life, that both thou thy may live:”

That thou mayst love the Lord thy God, and that thou mayst obey His voice, and that thou mayst cleave unto Him: for He *is* thy life and the length of thy days.

From this information so far, we learned the creation of life on to this Earth, from light, time, nature, Sun and Moon and all the armies of Heaven, animals, and ultimately all things created, and at the end of all creation: The LORD GOD created Man. All this was created by the Creator in six days.

The seventh day is dedicated to the memory for the people for rest and to be a sign between Men and God for ever and ever. And the first moment of the creation of life on this Earth, God created light through light illuminates the truth. Through truth he showed time and time for all those who have eyes or those who have time to find time and use that time for their salvation, in time is the revealed truth in which salvation is found which leads from Eden through time, through Jesus / Isaa, through the law of ten God’s commandments, faith and back to Eden. Since everything is created: everything visible and invisible is perfect and God created everything in perfect circle like Alpha and Omega, for in these circles of perfection, with God’s help and providence, the beginning is renewed, going in a circle to the end.

At the point where it ends, that's where it starts again. But everything has its time and its determination. And the length of his time, the Creator limited everything to his own time, the limit of perfection or the exact specified length for what is created and determined. And there is nothing created that is missing the purpose of its creation or the purpose for which it was created. All of these circles were, in fact, perfect creations of 360 degrees large or small, depending on their function, all have God's seal of Alpha and Omega.

And Christ / Isaa / Messiah says, of himself that he is the Alpha and the Omega. That he created everything and that only he maintains everything created in function and life and he will maintain everything until a certain time.

"And time, too, has its time."

The time after the birth of Christ / Messiah is called (the New Era or new period) and, that time continued to be counted immediately after the birth of Christ / Messiah and we count it even now. And since calendars have been changed in these 2.000 thousand years by human wishes, by these same human wishes were changed weeks as well as hours of the day. Time has taken shape and measure by human wishes. And God allowed time to be measured by the Moon and the Sun and every (day) from existence has 12 hours (night) and 12 hours (day) That means the middle of the day is at 6 am.- not 12 p.m. Although the number of hours has not changed, but has been (shifted) and turned in the opposite direction from (movement of) time. And that shift it changes time in dates and times that are very important to the truth which leads to eternal life.

The days have also been moved. The first day of the week was put on the seventh place. So the seventh day, which God had dedicated, came to the sixth place. The book of creation states that, God created Man on Friday, on the sixth, day. And on the seventh day, the Sabbath, blessed and consecrated it. There, too, because of the change of day of the week and their order, the most important truth that unites people with God is lost and which shows him and shows that he alone is God and that he created Heaven and Earth, as well as life on this Earth and the whole universe and that it is only in his hands and at his mercy our future life that we are all going to, some go to eternal death and some into eternal life. This truth was overshadowed by that modified calendar which was made according to human wishes and not according to God's instructions. The years also have changed the number of days. Perfect is the Creator everything was perfect so nothing changes so far, neither in the Sun nor in the Moon.

Everything works exactly as God created it. And that means that up to the birth of Jesus was exactly on the 4.000 years and he was born exactly at 4.000 years (Old Era). As the time shows exactly described in the holy books.

All the scriptures describe events that were important to establishing this truth, because this truth is of vital importance for each individual. All important events are accurately described, from

Adam / Adem to Noe / Noah, Abraham / Ibrahim, Moses / Muse, David / Davud and all the years they were born and the years how long people lived who had God-ordained the role of transmitting holy blood, pure without mixing with other blood until they transmitted it until the birth of Jesus / Isaa / Christ / Messiah. It shows that the time (ancient era was accurately measured). But time (New Era) has shifted and it is known when it was shifted, but if we count time since the world began, and we use weeks for that counting over which we count the time of 7 days or a month of 30 days, we will find that a year always has 360 days. Then we multiply, align with the days added to the years, time says that we have not yet reached 2.000 years. But if we have not reached 2.000 years yet, I will now not, for my reasons, write what year we are in. Though we still are close to 2.000 years. That speaks, by itself, that we are in the end of 6.000 years. That is the time when the time is running out that God gave to Earth and its inhabitants. Translated into ordinary man language, the time has come to the end of its Omega, it has accomplished its task and soon there will be no more time on this Earth, and that means nor life which we had until now. Because everything created goes in a circle of 360 ° and this Earth and life on it arrived at the point where the beginning and end merge or (Omega merges with Alpha). We received another piece of information, A (Alpha), is the beginning or birth of every creation that God created, and O = (Omega), is the end or the end of everything created, that God created. And together they make life and death and again life and death. Everything created it functions according to God's providence on that basis. And it is precisely because of us that the Creator created us, so that we can understand that after death there is life, and that we seek that life, that we find and enter it. All other creations of God are also born and die. New ones are always born so we too will be born in a different body structure because this body, which is of the Earth and which is dependent on the Earth and of its elements, changes into the body that Adam / Adem had at the beginning of creation. His body, though it was of the Earth, it had a different structure and was dependent on the food of Eden, that is the food that the Angels eat.

When they were cast out of Eden, God told them they would cultivate the land and that they will eat from it, and with the sweat of their faces. So after leaving Paradise they started eating the food they got from the Earth, which did not have all the nutrients on which life eternal depends and the days of their lives began to shorten.

From these above images, which the Universe itself shows, we will confirm that the information is correct and that it is for everyone available and that they are very understandable. Anyone who rejects their judgements will be able to see them, that is to whom God opens his eyes and God opens eyes to whom he wants. In the word of God it is written that the reasonable will understand. All this information, which the Lord God gives, is shown to us through his creations and his perfection. Everything created is perfect, but everything is the starting point (from the letter A); which means it is the beginning or Alpha, then going in a circle of perfect shape comes to end and merges with the starting point.

This circle that went around to assemble with point A, it made the glory of O in that tour, so Omega was created.

Omega means the border or a certain border to which it can reach God's creation around which the border or boundary went. That limit can be larger or smaller or very large, depending on the purpose God created it. If we start translating the words Alpha and Omega then we will come to this information:

A. First:

O. The Last:

A. First frequency of beginning:

O. Border:

The big (border) fences the genes in their original creation, after a while he connects them again with the beginning, with Alpha on which there is a restart in the same size and shape that is renewed from the first moment of creation and will be renewed until the time until it is necessary for that purpose to the Creator which he created it for this purpose.

Altogether, in its circle or border, it functions in its original state.

Omega has many different elements in every created element created in (Alpha and Omega) and they all make up life together of one particular entity. Each unit has its own important determinant task in its functioning and every one of these entities is renewed to its particular time and to its particular function.

Each specific entity fits into its exact specific location, given to it by the Creator at creation. So everything has its own Alpha and Omega, but all Alpha and Omega together are a whole which were created. Everything visible and invisible has its own 360 °

circle and its own Alpha and Omega. Nature has its own, the plant world has its own, the animal world has its own, and humans have their Alpha and Omega.

Everything created by God it was created through Alpha and Omega, through Jesus / Christ / Messiah, he is Alpha and Omega. In book of, (Genesis) writes, God created Man in his own form, and breathed his life's breath into him and became a Man, a living soul.

Man is created on the principle of Alpha and Omega and also has boundaries in shape, in skin colour, in character, it functions like all nature and everything that God has created, to last, to live, to die, and function on the principle of Alpha and Omega, by the function of God, in its limit, in its element from its composed Alpha and Omega, jointly united functions of various constituent substances and different readable substances that sustain a person's life as well as his length of life.

All these elements, which are found as the creation of everything created in its frequency substance, starting from point A, ending at point O, together they form the Alpha and Omega of all creation.

Light has its Alpha and Omega in its limit that it has received from God.

Time has its Alpha and Omega which it received from God in its own the limit of existence.

Water has its Alpha and Omega in its limit of existence.

The air has its Alpha and Omega in its limit of existence.

Humans and animals have their Alpha and Omega within its limits of existence.

Nature has its Alpha and Omega in its limit of existence.

The whole army of heaven has its Alpha and Omega within its limits of existence.

We get information that God created us and that we were created from 12 of the same elements called the foundations of life. These are the foundations of life based on 12 same genes, 12 minerals, 12 same chemical elements, 12 of the same colour, that has 12 diamonds.

We are connected in our creation with these same elements and we depend on these elements, these elements are the foundations by which the Universe and humans function.

Man is an integral part of the Universe and have main genes that are, also, part of the Universe and that we are all dependent on God who by force and his power he keeps everything alive and in function, until a certain time.

Because everything created shows its determined signs, because of which it is created, so the stars and planets, show their signs.

The Sun and the Moon have their own, God-ordained task: to show time through Days and Years.

And we need to use the time we go through, to see that there is a reason why we have been in our time and in our lives.

Time has its determined task. It shows and has always shown people that they are passers-by through time and take advantage of their own time, that they were not created on this earth for no reason.

To think if they are worthy to stand before their Creator, when the time comes to stand before him. What the book of their lives will show, which will come before God, in the day appointed for it.

Will they regret this passing time which they did not take advantage of, and were not honoured to stand before God, but with their shameful faces, as Adam / Adem, they will confess their sin and be expelled from the part of eternal life.

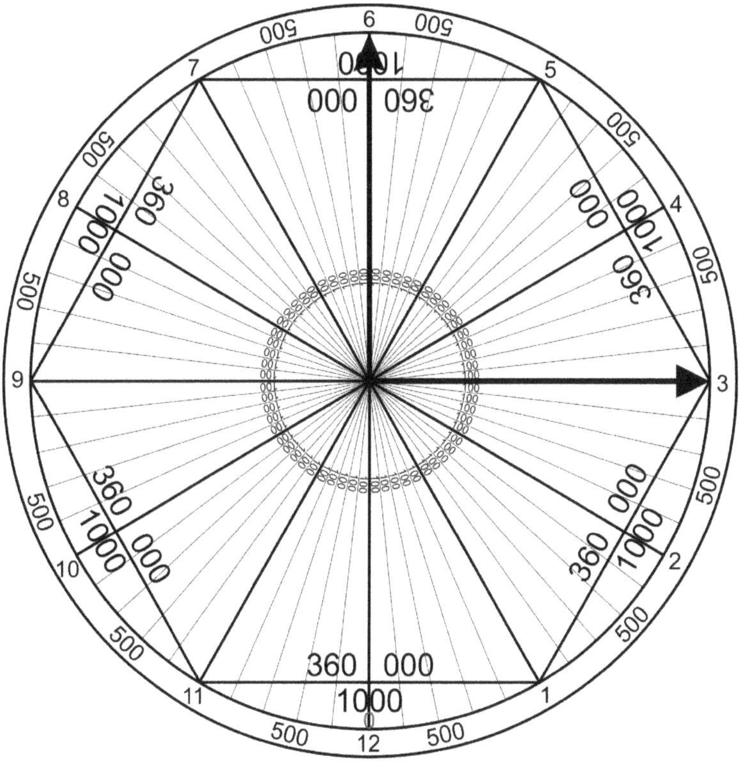

Picture 33 (GENES ARE THE BASIS OF THE LIFE OF THE UNIVERSE)
In this picture we see the appearance of the genes that are inherited and that are the foundation of everything the LORD GOD has created. They are at the foundation of the life of our galaxy as well as in every creature in nature, in the stars and in to all that is seen and not seen.

All these creations of God show the same shapes that are circles of 360° which speak of the perfection found in every creation which shows one and the same goal: everything is born and dies and is reborn in of the same dimension and shape, but still as a new creation. Such as people created on the same principle, even if they were born and died from before 6.000 thousand years ago, the same genes are always born, only in a new creation that has a slightly altered appearance. That life goes in its own direction and comes to the same starting point from where it started. That happens with all the creations of God that make up life on this planet Earth and without those many creations that are, in fact, one whole but divided into its God-ordained functions to create from their small contribution the whole of God's perfection and the line based on Alpha and Omega. And on that one way, they maintained a perfect whole on this Earth as well as throughout to the universe.

In these images we see the genes that are an integral part of the Universe and people, like we are all one whole of God, which God holds in his hands and in all his power. Everything created complements itself in its existence and everything created depends on God's power and God's wisdom. That power of God is in the middle it is the centre from which everything is created and to which everything created leans on.

It is a cornerstone which the MASONS (rejected) because they wanted to build a building of their own free will, the one that suits them. But without a stone, which is also a cornerstone which no building can exist, even if it exists, it will not last long because only that what is built on hard stone can stand. So we see that the cornerstone is also the Jasper stone on which the time, light and space leans whose name is Jesus/Issa; Christ/ Messiah. Through the centre of the circle, which is Alpha and Omega, which is Jesus /Christ. Isaa /Christ- Messiah, passed all the lines and all of them lean on him and through his strength and power they can exist.

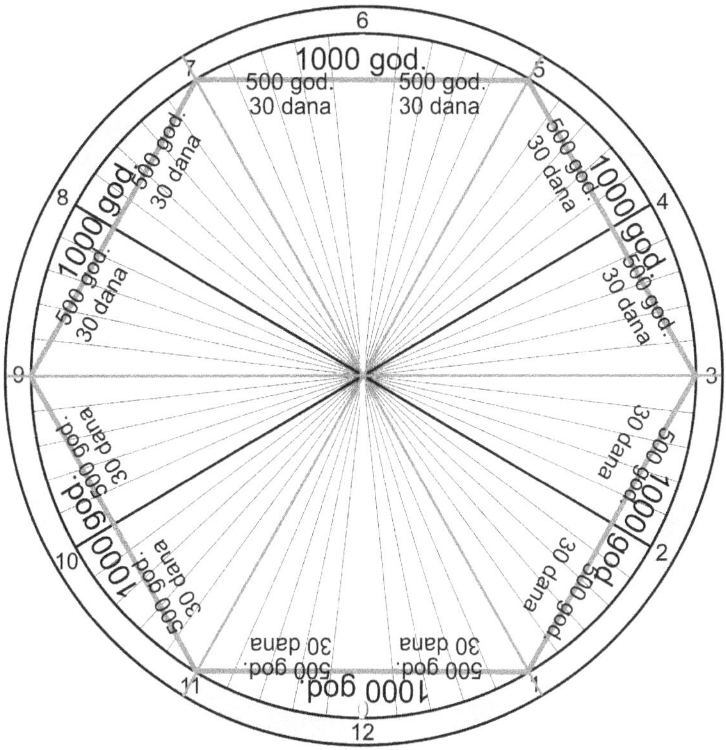

Picture 34 (GENES IN TIME AND SPACE).

It shows a 360 ° circle, then a hexagon. When the lines are drawn through time and space all forms of genes are obtained which in general are the ones who make up the whole of life, of the universe as well as of man. And Man is a part of the Universe (and has) the form of God.

Only man created God with his hands and breathed his life's breath and man became a living soul.
We get the information that Man was created by God from two different types of genes or in two dimensions of two different types of genes. Man, although part of the Universe, is also part of God who created him and who gave him two dimensions, i.e. gave him a body that is dependent on all elements and genes that sustain the life of the entire Universe. And God gave him both his genes and his body shape. But this form of the body, which we carry, also has a two-dimension: one dimension that is seen and disappears when a human returns to the Earth, from which God created him, and another dimension which is in us all the time and we couldn't see with the naked eye. Our eyes only see what is created of the Earth and its elements.
In fact, this second spiritual dimension is us and our being takes the form of God.
God is also a spirit and spiritual being and Angels are spiritual beings and exist in their dimension.
When the kingdom of God is restored to this Earth and all Men which Jesus / Isaa awakens to eternal life, everyone will have a spiritual form, it will be in spiritual form.
The first type of genes, which in their form and composition belong to all types of creations, they are called fundamental genes because they have all the creations of God, be they in matter that is included in living matter, which has soul and blood, or were in matter which is counted in matter which is alive but without soul and blood, i.e. in the form that the Universe has.

The universe is all living in its form of life and that form exists for the needs given to it by the Creator

And all forms of the genes also have a 360 ° circle shape, or clock shape by which time is also counted. The shape of space, the shape of time, they all find themselves within that circle in their Alpha and Omega and connect every God's creation in its life existence.

All live and exist by giving each other a part of themselves and taking each part of the other, through the paths of energy, electricity, magnetic forces and chemical elements, which God has set up to connect and unite in their existence. And it all takes strength, and power from the source of life and from the tree of life from the centre forces, the power of God. They transfer that strength and power to all creation.

And for the sake of Man, whom God created and separated from all others, he descended to Earth and gave his life in exchange for eternal life which Adam / Adem lost, to raise him to the heights where he was created, in sinless and eternal life.

These forms of genes always go on indefinitely depending on creations in which they exist for precisely defined purposes for what God created it.

All these entities show and are shown at the first moment of the creation of life on this Earth and are described in the first week of creation; when we start from zero; when there was no time, from the light that created time; seconds, minutes, days, weeks, months, years and thousands of years, then we see the path that takes us through time and brings us to the end of time, to 6.000 thousand years.

And that is the time that is counted from Adam's fall into sin and all the way to the end.

Because time constantly directs us for the hour of time then on that clock says that we have two times, the time that has passed and which goes away from us, and it counts as left time and goes from left to the right, and the second time; that comes to meet us is our time in which we live, which we are to follow, it goes from right to left and it's right time. Both times confirm the accurately the information, the past tense says how much time has passed since the origin of life on this Earth and this galaxy and how much longer there is until end to life on this Earth and this galaxy, in this the form we live in.

As the previous information explained that the week has 7 days and that God created everything in 6 days, and he dedicate the seventh day, and he put a sign between himself and the people.

The order of the days of the week:

Day 1 is Sunday:

Day 2 is Monday:

Day 3 is Tuesday:

Day 4 is Wednesday (which is the middle of the week):

Day 5 is Thursday:

Day 6 is Friday: In the sixth day God created man.

Day 7 is Saturday: Is a holy day and nonworking day of the week.

So this information says that Wednesday is the middle day of the week, middle of the day and night, middle of the year, and middle of the 6.000 years.

Just at that place, in the middle of the day, at noon, Jesus / Isaa was in mortal agony and just on Friday the same day when LORD GOD created man. (Adam / Adem).

This is confirmed by the Universe, as time goes on, that Adam / Adem was created on the day Friday, in the 144th hour, of creation life on this Earth.

Jesus / Isaa / Messiah gave his life on Friday, on the 6th day of the week and on 1.440.000 holy hour from the(day) since Adam's / Adem's fall into sin, then at 1.440.000 (holy) hour. Holy hours are hours that are counted and come as the sum of all the Sabbaths or (holy days) in the 4.000 years that have passed since Adam to Christ / Messiah. 4.000 years also indicate Day 4 of creation which God created Wednesday. It was at this point that the Sun was created that replaced the light of God and since then serves this planet and life on it. At that point, at the zenith, in the centre where the centre is counted in days or 6 o'clock at which the head of our Saviour and Redeemer was bent down for our sins. He endured the torment he did not deserve. Confirmed through the time that has passed and the time that will come, a time that will not be, that he is the one who created everything, who holds everything in life and in his own hands he maintains everything by his grace and love.

And he will maintain it until the appointed time. The universe and everything created knows its time. And that time, also, it shows to the people too.

Everything that God has created, he has created to serve men and to give his own signs that people do not get attached to the earthly beauty that strongly in which they are passing by but rather believe in the promised word of God as in events that confirm everything created around us, that we have God who created us, who confirmed himself in his own words and his nature, in his genes, which through time and through light, tells the truth so that no one can say that he was not instructed, and no one will have an excuse for their (disrespect) for God and his law.

His law respects everything created except the Man, for whom this was all created.

He put everything in the hands of Adam / Adem and said to him, "Everything created on the Earth is under your control, use, keep and take care of everything."

In the end, he gave his firstborn son in heaven to replace the firstborn son on Earth and he gave his life for every Adam / Adem's child, that no one dies who believes in him but has eternal life.

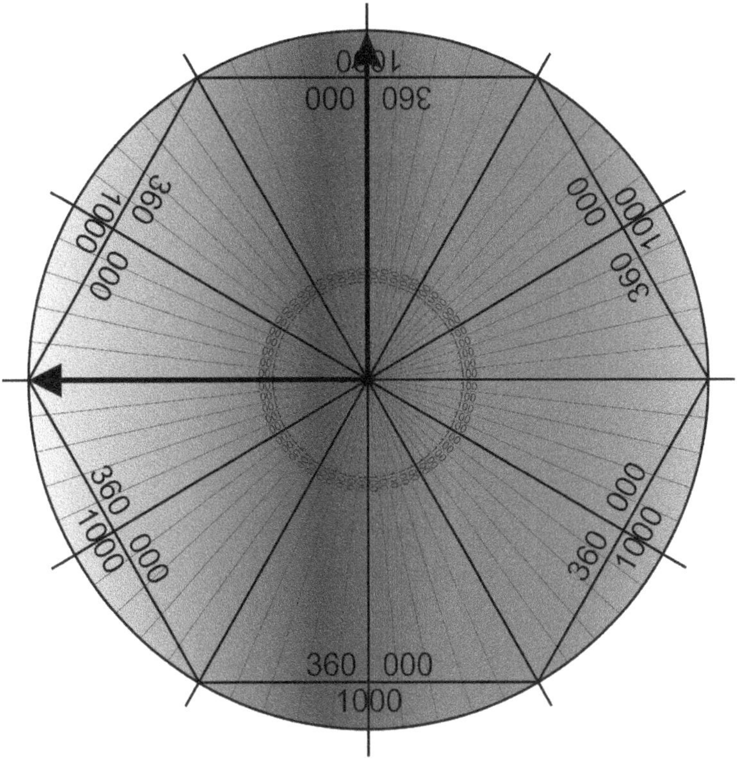

Picture 35 (RAINBOW OF GRACE EMBRACES THE EARTH).

The whole Universe is showing by its signs the direction we should take to God, the God of love who gave us another chance to let us return to the Garden of Eden, through the law of God and through the Lamb slaughtered since , from the foundation of the world. Through, Christ - the Messiah.

Nature says in her snowflakes - six.

Bees speak on their hive / in wax- six.

Saturn speaks on its body, on which is imprinted a hexagon - six.

Our genes say - six.

Time tells us - six.

Half every day has on the - six hours.

Half a year has - six months.

This is how nature tells us and informs us and shows us the way of truth and life, for all the peoples of the all language and tribe. Everything says that there is only one God, who created everything.

That is all and all truth about him, that a sinful man would return from his sinful path and through the path of truth and life he reaches the destination for which he was created.

God did not create Man to have a limited life and a difficult life. It was the free will of the first man who put his desires in front of God who wanted to have his own desires and to tempt God through his desires. He fell in love with Eve and because of love for her he left the love of God and law of God.

Everything created on this Earth tells us something very important, time has revealed important information to us through the light, all the numbers like all letters speak and direct us to another piece of information, time counts from the first second in which God created light and so through light God created time and in time gave us the first seconds, minutes, hours and years.

Going through a space that is bounded and that also has its Alpha and Omega, which also has 360 °degrees, and that the space is composed of their spaces from Alpha and Omega and so together they form one determined size that has its limit, its purpose, for what God created it be.

Space, like time, fits into a perfect 360 ° circle, just like 360 days a year or 360 minutes in six hours, or 360 seconds in 6 minutes. Everything is a perfect creation, a perfect Creator, a God who revealed, appeared and confirmed himself, through Abraham / Ibrahim, Isaac, Jacob /

Jacob through Moses / Moses, David / David who were obedient and faithful patriarchs and prophets of God and who, all in their time, carried the word of God and kept the word of God.

This confirms the time in time that the prophecy was fulfilled exactly on a certain day and time, which said for 4.000 years that the lamb of God would come which was slaughtered, from the beginning of the world or from the fall into sin of the first man Adema / Adem.

And which showed that the wages of sin is death; but since Adam/ Adem did not die, the day he sinned; the sacrificial victim, through the lamb, it was said, that the lamb of God will take away the sins of the world and that the fallen human race will get another chance.

Through the law of God given to Moses / Musa, on Mount Sinai and through blood and sacrifice of the Holy Lamb, Jesus / Christ / Messiah it will be possible to return back to the lost Paradise / heaven or eternal life.

Because he shows everything created, through time and through perfection of everything created in the form of a 360 ° circle and in that message we see the way through which to pass, number 3 and number 6 carry a very important message. When we add the number 3 + 6 we get the result 9. The number 9 has its final meaning and up to the number nine are the roots of all numbers, and beyond number 9, all the numbers are just repeated and in that repetition they show their God-ordained information. Through the numbers we have information by which we calculate time as well as space or numbers 3 and 6 make 360 as a perfect number and Alpha and Omega in space which it shows us the very key to the life of all creation. Everything created complements and functions in one great whole, in God, and through God, as the basis of beginning and end and beginning again.

It is the principle of Alpha and Omega as well as the message contained in it.

The number 360, which makes up the number 9 or 90 or 900 or 9.000 have their own message: 3 + 6 + 0 = 360

In number 3, number 6 and number 9 is the key to the universe. These are the numbers through which God has revealed the truth to us since the beginning of life on this planet Earth and to the very end of life on this earth.

So far we have had numbers that spoke, in detail about themselves from the very beginning of the creation of life through space and through time, through numbers which showed the days of creation and the exact length of time of creation of life on this planet Earth. All numbers used were:

0, 1, 2, 3, 4, 5, 6, 7, 8, 9, the numbers behind them just repeat. These numbers are the roots of all possible numbers.

The number 9 is the number where the root of the numbers ends. Number 9 is number which says that something has been fulfilled in that number and that in that number it will be fullness of all creation.

Such a repetition of the number 9, which we get in every Alpha and Omega.
When we look at each circle, whether large or small it has 360 °.
In 360 ° there are numbers 3 and 6.
3 + 6 = 9, which means that each Alpha and Omega are one separate nine.
When we divide this circle into 2 we will get: 360: 2 = 180 or 180: 2 = 90 or 90: 2 = 45 or 45: 2 = 22.5. When we add up each of these numbers the sum is 9.
Since zero in reality has no value, but serves as a seal for confirmation of accuracy. They show the time in years, such as at 100 or in 1.000 years, the time that has already passed, and by zeros we also determine the accuracy of the years which those zeros show.
Eg. 1 year has 360 days - there is only one zero here, but 4.000 years has also three zeros + 4, so we read that there were 4.000 thousand years. These zeros serve for time orientation and for a clear understanding of the time the information speaks for.
In reality, zero has no value, so we see that Jesus / Issa died exactly on the same day and time that Adam / Adem was created and that he was a 100% substitute for his sin and his death.
From this information we can see that all these numbers fit into the number 9.
Like 144 when we divide separately, we get the number 9: 1 + 4 + 4 = 9.
When we add up the numbers 1 + 2 + 3 + 4 + 5 + 6 + 7 + 8 = 36.
These collected numbers, up to the number 8, say that the sum is 36 and the number 36 so far keeps telling us that in numbers

3 and 6 or 36 stands the secret of all secrets. It was at number 8 that he discovered another secret.

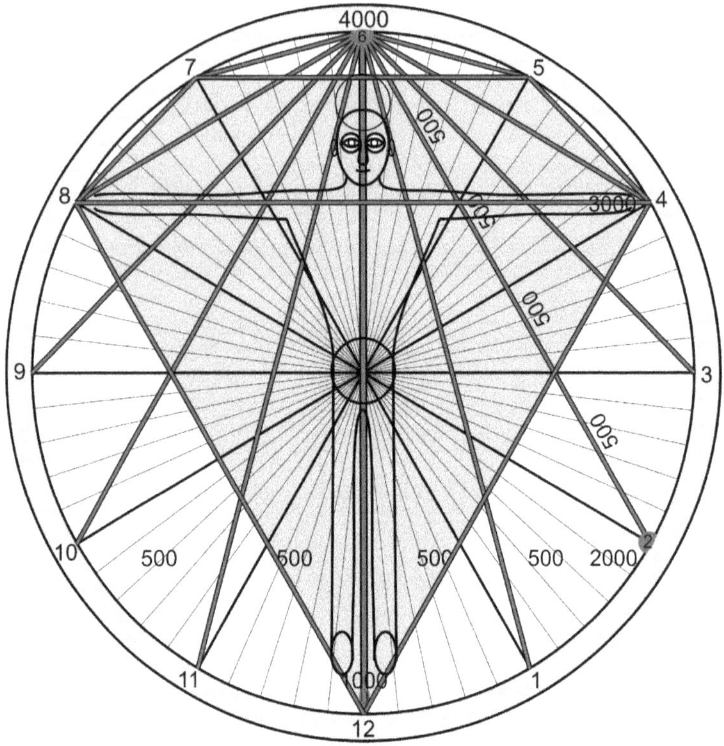

Picture 36 (STONE JASPIS / ALEM THROUGH TIME AND SPACE)

Shows the time, which shows first minute then hours, or the current time we are in and in which we currently live. It then shows the thousands of years in which these events took place.

In this picture we see the Creator, Redeemer and Saviour for which was spoken about in the past tense, for which the present speaks, and for which the future speaks. In this picture, the Universe itself shows us and confirms who created all this, who holds and keeps it in his own hands, who gave his life for our sin and who will come and revive us at the appointed time, which our Lord God has appointed for us our Father in heaven or our merciful Creator. This picture, through time, space and light in the middle shows a cross on which the plan of salvation would be fulfilled and on which Jesus / Christ would die so that we could live.

Here, at that particular predicted time, through the Sun and the Moon, which is used to count our time, we found God's lamb slaughtered since the beginning of the world, we have all come to our own (lamb) which is slaughtered for every individual because his life has so such a price that he can redeem anyone he wants, with his own by the will of his will, to acknowledge that sacrifice made for him, and to seek from the Lord God through this sacrifice and forgiveness through him inherited and acquired sins.

We see a 360 ° circle and time in it. We see the light that illuminates the truth because first God created light and through it we get light - day.

In the light we see that at exactly 4.000, right in the middle of the week at the place where God created the Sun, in the middle of the day at noon, on the 6th hour, at 1.440.000 a day that fits in 4.000 years, because one year has 360 days x 1000 years = 360.000 days, and 360.000 days x 4.000 years = 1.440.000 days, on that day the prophecy was fulfilled, and on that day the Creator, Redeemer and Saviour was born.

This is where the counting of the years we count in (the Old Era) ends. On that one time draws a line and divides two time periods, just as they divided day and night, truth from lies, eternal life from eternal death.

This is where the truths of all time are revealed: past, present, and future.

After 10.000 Weeks of Genesis, from Adam / Adem and humanity on this Earth, one week has (6) six working days or 144 working hours.

10 weeks has 60 working days or 10 x 144 = 1.440 working hours.

1.00 week has 600 working days or 1.00 x 144 = 14.400 working hours.

1.000 weeks has 6.000 working days or 1.000 x 144 = 144.000 hours.

10.000 weeks has 60.000 working days or 10.000 x 144 = 1.440.000 hours.

This information also confirms the specific hours at which born was Jesus / Messiah.

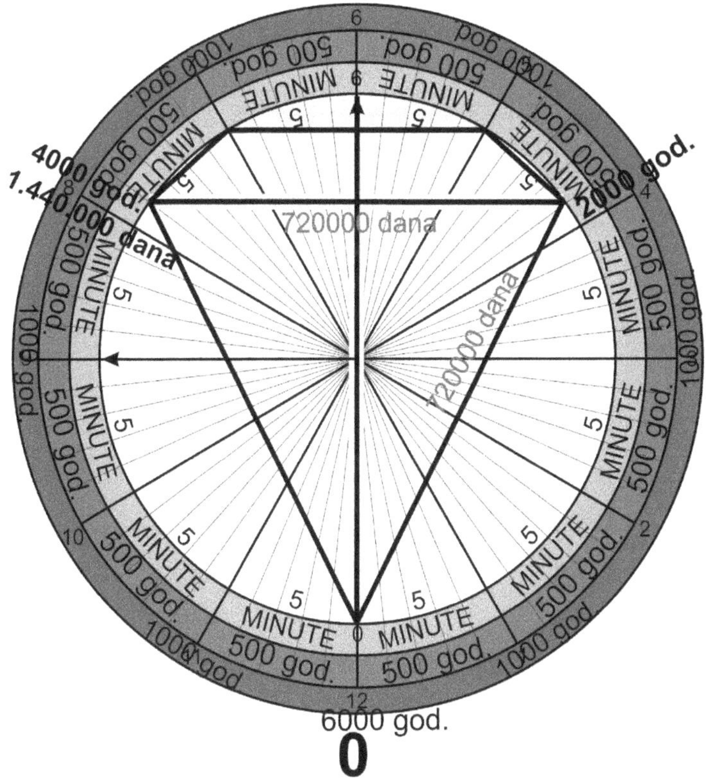

Picture 36a (ALL ALPHA AND OMEGA ON STONE JASPIS / ALEM).

Other information that also comes along with in this direction indicates that it is from zero to 6 o'clock, where the light went through the darkness and divided the darkness from the day, in parallel with the number 8 according to number 4. And vertically fits with number 12 and number 6.

This location shows hour 8 and hour 4 horizontally + hour 12 and hour 6 vertically. We see the cross on which he perished, and our redeemer died on a specific day time.

On a particular year, the exact day and time, on Friday, on the day when the first Man was created who sinned, also because of his sin, the Creator himself came to pay the price of his sins and his death, with his own life. Through his death, he enabled a sinful Man again a return to eternal life.

Number 8 has its information: all numbers from 1 to 8; together form one whole: 1 + 2 + 3 + 4 + 5 + 6 + 7 + 8 = 36, that is the time when the torture of Christ / the Messiah began on Golgotha, until his death, from dawn, when he was sentenced to death, until he is in death agony at 8 o'clock with his right hand, turned the KEY. A KEY that had double power, which is divided two times at the same time: the time of eternal death and a time of eternal life. At that place at number 8 there are also two circles two Alpha and Omega, because they are divided into two different times

---

In this place, through the death of Christ, by which it pays off Adam / Adem and our sin and where in his mortal agony it unlocks the door of eternal life at the same moment locks the door of eternal death, and through his sacrifice pays off sin for all people on this Earth, that no one who believes in him does not lose eternal life.

This lasted for 8 hours or collected through confirmation of time which from the beginning of time spoke and directed the generation of the time, to events that need to happen, and to recognize him as his Saviour and as his King of life.

These numbers through the hours, they spoke and confirmed the truth.
All of these numbers, which are also the roots of numbers and everything has its roots and their top, they say that at number 8, there will be an important event, explore which!
There Alpha with Omega merged , at number 36. And at number 9, the numbers get fullness of the execution of the plan of salvation: 1 + 2 + 3 + 4 + 5 + 6 + 7 + 8 = 36 + 9 = 45.
45 is again number 9. or 4 + 5 = 9.
90 is the middle and centre of the Garden of Eden / Paradise / Jennet.
4 x 90 = 360. In this centre is the tree of life.
Alpha and Omega, everything created in this galaxy complement and fulfill every creation of God. They all go in the same direction and show day, hour, month, year, thousands of years on which was born and died the Creator, Redeemer and Saviour. The one who has the power to get us to rise to a new life. This information says that at 8 o'clock the time of Christ's work on this earth was over and that the plan of salvation is completed. And at 9 o'clock the life of Christ was extinguished.
Through these numbers we also get the most important information and that is, what the numbers should show, which is that at number 9 the time of this Earth and the life on it in this form will be over.
The fulfillment of time, and the cessation of time or the end of time at number 9 which is also the ninth hour of the day or at year 9.000 which also ends all work on this Earth.

9.000 years is obtained when 3.000 years is added to 6.000, then we have exactly 9.000 years.

This number fits exactly into the perfect Alpha and Omega and shows that there is the first point or point A or point O or the starting point of life.

That is also the end point of Omega, which travelled time and space and completed its function. When the Mega is composed with Alpha, then the period and life that took place in that creation ends and is renewed in a new creation. Which would mean: this form of life ends on this Earth because time has passed to all creation as well as to time itself.

That is true at number 9 or on 9.000 thousand years.

Or 6.000 years from the day Adam / Adem began to live on this Earth as a sinner, + 3.000 years of Adam's / Adam's life in the Garden of Eden.

That's a total of 9.000 years + 1.000 years that didn't count working days of the week.

That 1.000 years are counted in the holy Sabbath days.

This information says that on the eighth hour of the day, fulfilled the plan of salvation and that it is on that hour, on which are also (two) Alphas and Omega, or two 360-degree circles and they confirm the truth of the word God. Because only number 8 has two circles or (two) Alpha and Omega merged the two times that Jesus Christ in his death separated.

It is a time of eternal death and a time of eternal life.

No one else could separate the eternal death, earned by Adam / Adem, to cancel and justify, except the one who is also the king of life and who overcame eternal death - with his eternal life, and who complete the plan of salvation.
In the plan of salvation, which the heavenly Father together with his own son, promises, that his firstborn son in heaven Jesus-Isaa, will be a replacement of his firstborn son on Earth, Adam /Adem, where Adam / Adem replaces his eternal life with his free will, for eternal death.
God out of his love for man whom he created; agreed that Hi should be (both – the victim and the lamb), through which he will save the human race from eternal death.
Throughout the entire period of 4.000 years, sacrificial lambs were slaughtered which were referred to a sacrifice that would come at a certain time and complete that plan of salvation. It happened at a specific time which was foreseen for that in the plan of salvation.
In time, through time and the Universe, it is confirmed in all dates and events that the information is accurate and that this information can come into the hands of every person who has requested such information.
In those torments of death, the KEY, which has been with him since the beginning of sin on, the right shoulder; locked the door of eternal death, and at the same time unlocked the door of eternal life.
In his mortal torments, the moment he died for this whole human race, at that moment by his death he paid for the sin of every one,

who seeks forgiveness of sins, through his death. In that hour, at 8 o'clock it was confirmed with two seals, in the moments when Christ / Messiah shared with life, these were also the moments in which completed the plan of salvation for which he came down to on this Earth.

At the eighth hour of the day , in the darkness that covered Christ / Messiah, from the sixth to the ninth hour of day, on Friday, Christ / Messiah died for every soul that believes in him. That through his sacrifice he too has eternal life.

And in the ninth hour of day, he let out his soul and gave his spirit to his Father. His disciples removed him from the cross before the beginning of the Sabbath and buried him in a tomb to which no one had yet been buried. The following information says: Jesus / Isaa / Messiah exactly on Friday, before the start of Saturday, was prepared for burial and buried in the grave, at a time before the Sun has yet set. This information confirms the a new day begins the moment the Sun sets. And confirms God's saying; "And it would be evening and it would be morning; the first day. "

In the book "New Testament or" Ingill "it says, Christ / Messiah lied in the grave Saturday, (day 7). He laid on the holy day in the tomb: Then again would be evening, and Sunday began, he was in the grave. But on Sunday morning he was not in the grave. In the book "Ingill" or the New Testament, it is written that on the grave women came and saw the stone rolled away. The Roman guard was at the tomb but they fainted at Christ's resurrection from the dead, and that the witnesses saw only the cloth in which he was wrapped, nicely put together. But two angels appeared to the women and later to his disciples, asking them who they were looking for.

Information that confirms the four holy books written by Christ's disciples, the Romans in their writings, and in the Greek and Hebrew writings, which confirm that Jesus / Christ / Messiah lay in the tomb for 36 hours, i.e. Saturday night and Saturday day; and Sunday night.

Christ would be on the cross and be tortured for 9 hours but at 8 o'clock with the eternal KEY he locked the eternal death, and the one who caused it,(Satan) his lie, and his darkness, which he created with his lies. When we add up the hours of Jesus / Messiah tortures, from the first hour, starting: 1 + 2 + 3 + 4 + 5 + 6 + 7 + 8 = 36 hours, that time shows how God showed, from the moment Adam/ Adem committed sin, and as he described in the plan of salvation, that the one who has the KEY, on his right shoulder or on his right hand, will come. He who has power over death that through his death eternal life will be born. And that at the moment of his death, Jesus / Isaa / Messiah locked eternal death which Adam / Adem unlocked with his sin.

There is no other name under heaven that can save Man from eternal death except the only one God and Jesus Christ who gave his life for us.

Jesus / Messiah was in the tomb for exactly 36 hours. This information says and confirms all other information, which say nothing else through these 6.000 years and which have only one point in the Universe to illuminate and to show it. And that is the fulfillment of the plan of salvation through Jesus / Messiah.

There is no other name under heaven that can save Man from eternal death except the only God and Jesus Christ who gave his life for us.

In the sixth day, when God created man, he made him in his own form, with his hands, and everything else created by the Lord God would be created only by the word "be" and it was. He gave Man his form and his genes, as he gave to all created basic genes and genes that separate species from species and genes that separate nature by its kind. Each species has its own basic genes that are the same for everyone. These genes are the foundation of life through which everything exists. They were divided by species, each species has its own fundamental genes that are renewed in its own species as a perfect frequency along a perfect line of 360 ° degree of Alpha and Omega.

Man has a special place in this universe, and because God, created all this for the benefit of Man, then Man himself has value in God, and that's why she was such a victim. Because of God's love for Man, it would be a pity for him to die and to disappear. Everything created by God is holy, but people do not even notice that they behave in this way towards God's creations.

They lost sight of the fact that God created all people and that he loves us all equally, that no one's sigh or tear will go uncollected if it is caused by someone else's arrogance and for no reason. All those sighs and tears will meet every person who challenged them and who in his lifetime did not see their unworthy conduct and who, while there was still time for repentance, did not seek forgiveness from the person they had harmed and reconciled or straightened the line they had moved and descended from their sin from God's way. Likewise, if they asked God for forgiveness from the person to whom they had inflicted grief, tears, or material damage then through forgiveness before God and the Angels, the angels who record that event, reconciliation and apology that sin before God is erased and people. That sin disappears from the book of life, and the book of life will be for each individual to open on the day we stand before our Lord, on the Day of Judgment. This means that every sin that is not forgiven before our death goes before us, we will be held accountable for it. No one can ask for the forgiveness of our sins for us.

Only he who has committed sin has a chance, while he is alive, to correct his sin, if God gives him wisdom from his wisdom and grace from his grace, if he opens his spiritual eyes and gives him the opportunity to realize that there is nothing on this Earth so valuable that we should lose our eternal life and the grace of God which warns him that we are passers-by through our own time. The time used by each individual serves us to see through that borrowed time, and through God's instructions where we need to go and which is the direction of eternal life and eternal time.

And God who is equally to all the heavenly Father and sees everything and everyone. Everyone on this Earth, that is still alive will see the results of their behaviour, whether he is positive or negative, before his time passes even before he falls asleep in a hard sleep he will see the result of his life's work and will begin to reap the fruits of his labour on this Earth. All his results will not be paid only to him but it will be paid to his successors as well.

There is no person on this Earth or money to bribe God, nor by which the forgiveness of sins can be paid. Everything earned is going in front of us and everything earned comes right in front of us at the moment when his time of God's patience ceases. And what which will come, it was our free will in which we lived according to our understanding and our conduct in relation to God and in attitude towards people, i.e. everything that we did wrong we will pay for it in this life. And all that is not we will be fully paid by our successors, our children and our grandchildren, until the fourth generation, until our sin is atoned for. Everything we have sinned against God, we will receive in person in the heavenly court and every unrepentant sin before Men and before God which is not erased from the book of life will testify against us and will take away our right to eternal life.

The wage of sin is death, and all the sins that come before us on the Day of Judgment day show in the book of life that they are un repented of, are only death.

Until the period that belonged to the (Old Era), people were justified through

faith in the promise that the Creator will justify us through sacrifice- the (Curban) Through the one who was abaut to be born and to die at the exact time and at the exact day and hour.

The time in which we live is called (the New Era), and is counted from the day of the birth of Christ / the Messiah, and the time of the birth of Christ / the Messiah and his death is shown by the entire Universe.

It shows the light, it shows the hour and in the hour we find the past, present and future time. He showed us an hour, the day and the hour when God created Man, showed us when man sinned, he showed us the time that has passed since Man's fall into sin, showed us the same time the fulfillment of the plan of salvation. He showed us the sacrifice/curban that God offered for the salvation of the sinful Man.

And every punishment we need to receive, we will receive one part on this Earth and the other part of life that we go into. All that we have done, positively or negatively, we leave it to our loved ones to inherit. If our descendants were evil until the third or fourth generations they will pay for our sin.

If we lived a life of faith in God, we were careful not to tread God's law, the 10 commandments of God and depending on what we believed in and to whom we have prayed to, we will receive a reward and a blessing still on this Earth. That reward for obedience and faith through the word of god brings blessings to all still on this Earth and the second part is in new life. A reward that leads to eternal life and a reward through the blessings that our descendants inherit.

If we believed that we were children of God and if we found a way in that belief and followed the path of established facts, we will have open eyes, we will see that we are the truly children of God and confirm, the truth will be before our eyes, as well as before our spiritual eyes which see beyond these eyes.

If we stand in front of a mirror and raise our hands to our shoulders horizontally, we will see that we are (the cross) on which our Creator was crucified, because of us and our sins.

All of us through Adam / Adem inherited sin and none of ours good work cannot save us from eternal death. Not even the Angels could give their lives for our sins, but rather Christ/ Messiah the one who created us, his sacrifice alone was enough to wash us from sin and eternal death. That is why he is our Creator and at the same time our Saviour and our Redeemer and our only possibility to justify his blood and his sacrifice and cleanse our sin. Only his sacrifice and his blood could protect us from eternal death. Anyone who believes in him and who obeys the law of God, 10 commandments of God and keeps them, gets the right to eternal life through God's love and mercy.

But those who did not know and who lived as they lived, and they have a second chance through Jesus / Isaa / the Messiah and his sacrifice, keeping the law of God receives the right to eternal life. In the law are all the instructions to the right life. In Christ / Messiah: is life, because he is the one who holds the life of everything in his hands and who gave life to everything visible and invisible, who is called (tree of life) and who is called (water of life). When we add up the numbers: 1 + 2 + 3 + 4 + 5 + 6 + 7 + 8 we get the number 36, and the number 36 shows the time in the hour in which the Messiah / Christ was dying. It was a time and a number in which death lost its victory over a sinful man.

On that eighth hour 8, in the darkness that enveloped Christ / the Messiah in his last hour of life, when he was dying for Man and humanity, he won victory over eternal death, over lies over the greatest enemy of God and Man.

Here our Creator, our Redeemer and our Saviour defeated Satan, the old serpent and the dragon who for his own benefit deceive Man, created by God and with his lies occupied this world and the people on it and makes false promises to people through his lies. Satan is the father of lies, the first lie was uttered by Satan himself, the only culprit for so many tormenting people on this Earth. God who is in heaven out of his love for Man, forbid the taste of the fruit from the tree of life. After Man's fall into sin, for that Adam / Adem had tasted the fruit of the tree of life, after God's commandment by doing what was forbidden to him, Man would live forever in this form of life, as we live now, in sickness, in wars, in injustice, that would be the life on Earth.

God, who is in heaven, made the greatest sacrifice that could be made for this human race, create a plan of salvation, gave his firstborn and the first created son in sacrifice on Golgotha and save us; through his sacrifice in his plan of salvation, deliver us from eternal torment and that each individual has the right to eternal life of his own free will and by using the truth that reaches him.

When we add up the numbers: 2 + 4 + 6 + 8+ 10 + 12 + 14 + 16 + 18, all the numbers show the same result: 90;

3 + 6 = 9 or this other sum 9.

What does God want to tell us? What is so important in number 9?

We will see what the Universe says, sending God-given messages to us for understanding the truth as well as for confirming that truth. Since in numbers we find this truth then we go through God's messages inscribed through numbers in time and space.

Here is the beginning, the first second God created and through which he began to count time. The Creator of all things and the Creator of time himself says, "Every night and every day together they have 24 hours."

In 6 six days, the Creator created everything, this planet of ours The galaxy, and the (7) seventh day is dedicated and separated from these (6) six days, in which he created everything. At 7 the seventh day God did not create anything because he created all in those 6 six working days. He created Man, last at 6 the sixth day, on Friday, and the seventh (7) day is the Sabbath which is the sign and confirmation that there is only one Creator and that he is the only God, glory belongs to him and thanks for all creation. Here we can conclude that we don't need to thank anyone either to give praise and glory other than the only Lord God who said to Jacob / Jacub: "My name is (Lord-God), but I am revealed to Abraham / Ibrahim as (Almighty God)."

This information says that the Creator also revealed in these messages his name as well as every secret we need for salvation. He revealed it at the exact time, which was needed by that generation. Every time has its time, but some secrets were revealed many years earlier because it was necessary for that generation, because of the instructions given to them and because of their salvation. Through the time that has passed we have received the greatest information and we have been shown everything that has happened on this Earth, and everything is recorded and written. So we have all the information that existed and which we shall yet get.

Time and counting time shows : That they have 24 hours every day and night .

One week has 144 working hours + 24 holy hours = 168 hours:

10 week has 1.440 working hours + 240 holy hours = 1.680 hours:

100 weeks has 14.400 working hours + 2.400 holy hours = 16.800 hours:

1.000 weeks has 144.000 working hours + 24.000 Holy Hours = 168.000 hours:

10.000 weeks has 1.440.000 working hours + 240.000 Holy Hours = 1.680.000 hours.

Or:

10.000 weeks has 1.440.000 working hours:

10.000 weeks has 240.000 holy hours:

Or:

1 week has 6 working days + 1 holy day:

10 week has 60 working days + 10 holy days:

100 week has 600 working days + 100 holy days:

1.000 week has 6.000 working days + 1.000 holy days:

10.000 week have 60.000 working days + 10.000 holy days:

60.000 working days x 24 hours = 1.440.000 working hours in 10.000 week:

10.000 holy days x 24 hours = 240.000 holy hours in 10.000 week:

10.000 weeks x 24 hours = 240.000 holy hours x 6.000) years = 1.440.000 holy hours.

Or:

1 month has 30 days;

1 year has 12 months;

12 months has 12 x 30 days = 360 days.

Or:

1 year has 360 days:

10 years has 3.600 days:

100 years has 36.000 days:

1.000 years has 360.000 days or 360.000 x 4.000 years = 1.440.000 days.

From all this information we have only one piece of information, that our Saviour was born on that day and that hour, all the time pointed to his coming, its birth, and fulfillment of the plan of salvation, and redemption of this planet and its inhabitants from eternal death. The whole Universe only shows this because the. Lord God would pity the Man he created, to die an eternal death, and out of his love he found a substitute for his mortal sin, He gave the firstborn son whom he created in Heaven for the firstborn son whom, he created on Earth, that every individual has a chance through his sacrifice and through his life to be returned to where he was his, also created, in the Garden of Eden, Paradise.

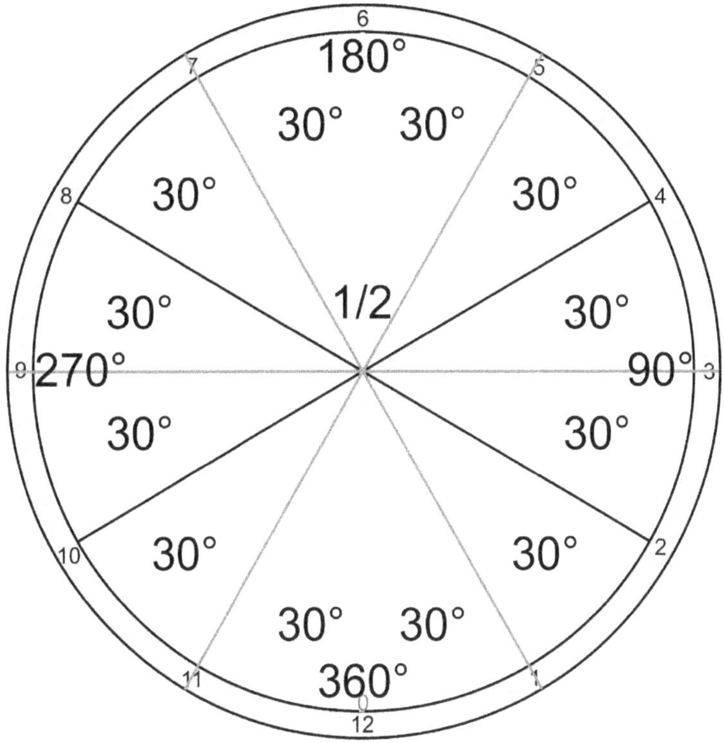

Picture 37 (SPACE AND TIME ON THE SAME LINE).

It says that there is also space, that its surface too the perfection of Alpha and Omega of 360 °; divided into its special parts that make up their wholes, God-ordained boundaries, which show that each part speaks for itself, and that is: one whole has 360 ° or 3 + 6 = 9.

One half has 180 ° or two halves together 360 °

or 1 + 8 = 9 or 3 + 6 = 9.

or 1/4 has 90 ° or 4 x 90 = 360 ° 4 x 9 = 36 = 3 + 6 = 9.

or 1/8 has 45 ° = 360

or 8 x 45 = 4 + 5 = 9.

Like all numbers, starting at zero where there was nothing, in order of 0, 1, 2, 3, 4, 5, 6, 7, 8, 9. Here is the end, after these numbers, all other numbers are just repeating. That means these are the roots of the numbers and they stopped at number 9, also in space as well as in time.

When we continue along the line of perfection we go through the numbers and time, they will take us in the direction we need to know. Because the Lord God who is in heaven, gave people every piece of information and explained everything through his holy word and holy books, through his own prophets and patriarchs, so we are not ignorant of his truth, so that we would not doubt the accuracy of his existence and instructed us in the accuracy of his name and of his throne.

Jesus / The Messiah said when, we pray we should say Our Father who art on heaven, and stressed the importance of those words. The one and only father has a throne in heaven. All other Gods to which people pray are on this Earth or in this galaxy.

Many people pray to nature, and to the Sun, and to the Moon or to the army of the heaven or to the statues, that people have made and call them Gods and pray to them for help. But those who want to seek God should seek him in heaven, because he created the Sun, the Moon, and all the host of heaven, how they are called, all the stars and everything we see in our galaxy. Going further, we come to this information: the almighty God, who is in heaven, he created everything in this galaxy of ours in 6 six days, and on (7) the seventh day he rested, and consecrated it. This also started counting the time of 1 second in which through light time was created.

Going through days, weeks, months, years and thousands of years, we go through time and the times:

From Adam / Adem to Noah / Nuh it is 1.000 years:

From Noah / Nuh to Abraham / Ibrahim is 1.000 years:

From Abraham / Ibrahim to Emperor David / Daud is 1.000 years:

From Emperor David / Daud to Jesus / Messiah / Isaa is 1.000 years:

That is 4.000 years since the creation of the day that counts (since) Adam / Adem left Paradise / Paradise / Garden of Eden for sin.

And they were counted like that until the plan of salvation was fulfilled and until the saviour Jesus / Isaa/ Messiah was born, exactly on 4.000 year. On that day, the time for salvation through the promise ceased to be counted. There the time we call the Old Era ends and the time of the New Age begins in which promise became a reality.

In the beginning, after the birth of Christ / the Messiah, when he finished for which he was born and descended on this Earth: to redeem all those who believe in him, through the truth they have received, through his life and through his death on this Earth, and through the truth that he returned to

heaven to his father and our father, fulfill the plan of salvation given to the first Man, through his love for Man and through his father's love for Man, he overcame sin and death.

When a person, through faith and prayer, in the name of Jesus Christ / Messiah, seeks from God the forgiveness of these sins and the purification that the heavenly father has given him the premise for life everlasting, on this Earth, that he may return again into the lost Garden of Eden.

After Christ's birth, death, and ascension into heaven, we continue to count the time of our (New Era). We are currently living in 2017 and times say 4.000 years of the Old Era and these 2017 years of the New Era together make 6.017 years from the day Adam left the Garden of Eden or sinless life.

Everything created shows 360, either years or degrees, and says it is the number 360 the number of perfection we have found so far in the Universe, showing Alpha and Omega, the perfect beginning and end, that merge and recreate the same creation in its new creation.

From this information we learn that there was a disturbance in the counting of the times in the (New Era).

Since the perfection of God is based on Alpha and Omega then a year (cannot have) 365 days nor 354 days, time is counted from the origin of time, from the first second by the Sun and by the Moon, how the Lord God also counted the first Week of creation: And it would be evening and it would be morning day first : day two: day three 3: etc.

That's how time (of the Old Era) was counted right, on Christ's/ Messiah's birthday because it is on the perfect day which is 1.440.000. Day from Adam's/Adam's (creation and leaving paradise). Jesus-Isaa was born.

Time that Adam / Adem spent in paradise is not counted or added at this time. That time is counted separately from the time in which Adam / Adem began to live after sin and in which we live today. It stayed recorded the day and hour in which the Lord God created Man, which was Friday, the sixth day of creation on the 144th. hour of creation.

When the time of these 2.017 years from this New Era is added up, when we go back to the time that was counted in the (Old Era), we will get a result that shows that we are not even in (2.000 year). When we subtract 5 days from each year, for all these years counted that way, we will get time according to God's calendar, not human, and then the result shows that we are not even in 2.000 years but we are very close.

The Light and the Universe have their information to reveal because it is the time for every revelation and there is nothing more hidden (than the day and hour) of Christ's second coming, when he will come on the clouds to raise from the graves his chosen ones. Who are his chosen ones? These are the ones who believed in him, believed that he was the lamb of God slaughtered from the beginning of the world because of our sins and because of our salvation.

The line of light and truth goes on and says; until Jesus / Christ / Messiah, 4.000 thousand years or 1.440.000 days, have passed. 2.000 years should pass from Jesus / Christ / Messiah to fill in 6.000

Why 6.000 years? Why is the whole Universe showing number 6?

On the (Saturn) we have a hexagonal seal showing number 6.

A snowflake, which falls when it is very cold, has six branches.

Accurately drawn number 6 on wax made by bees.

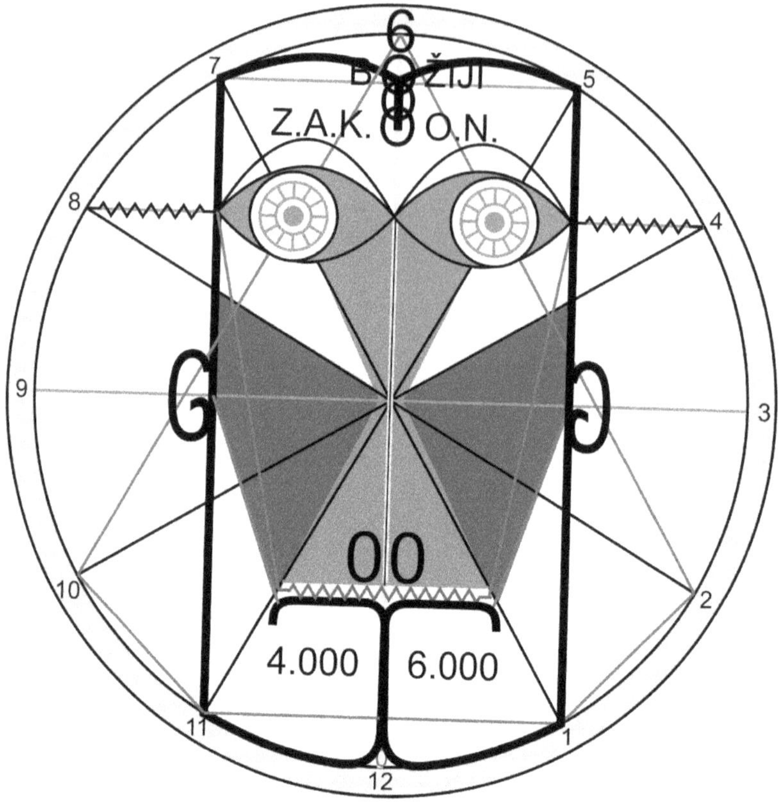

Picture 38 (UNIVERSE AND LIFE ON THE SAME LINE).

On our skin, on our face, on our body we have an inscribed number (6).

In this picture we see a 360 ° degree circle in which the face of a Man is located.

Lord God, showed us one very important information, our face and the shape of our body shows, in its direction of truth, that we have two stone slabs on our face or the law of God’s 10 commandments.

From number 7 to number 11 there is a vertical line. That's the line from the end forehead, ear to chin. He bends his chin to his lips and the line shows half of his face.

In the middle of the forehead, it bends its shape and separates the two halves of the face and makes the number six 6.

In the middle of the forehead and in the middle of the chin is exactly the middle of our face. With that same line that goes to the other side of our face, with left side, from number five 5, and vertically to number 1, there it bends and makes another stone tablet of the law of God.

On the right side of our face are the 4 commandments that apply to God.

On the left are the 6 commandments that apply to people. Above the blue line, which borders the lips, are two zeros which separate our nose and show the direction of movement, which is in blue colour. On each of these plates we have two fish. Because when these lines reach the tip of the nose they go across in the direction of the eyes and there they get the appearance of a fish, or an hour of time.

Right in our eyes we have the look of Alpha and Omega and in the middle, where pupils are located, we see an hour, which showed and confirmed all this time, which we had and which will come.

Two times also appeared before our eyes. The time of the Old Era and a time of a New Era, which makes 6.000 years. The fish show the form of genes that are in the eyes and that continue indefinitely, in another time to come. Our ears have the shape of number 6 and confirm, too, that through 6 days came 6.000 years and that the number 6 showed all the Alpha and Omega that constituted this form of life, that these were all the days or years on which all work connected with this world was completed, and that behind every number six 6 comes, the next number 7. The number seven 7 is also a sacred number on which the working and holy hours or years and thousands of years are composed.

All of this happens through the time in which we live and through this time we will gain eternal time. In this picture of faces we see many types of genes that pass through our face and who through Alpha and Omega continue their direction to eternal life.

The middle of the day and night are at number 6.

The universe says 360 or 36, when we put all this on the line of Universal direction and perfection, we see the direction of perfection or number 36.

When we add up the days of a New Era that ends at 2.000 years, the result says that 2.000 years have 720.000 days.

When 720.000 days are added to the 1.440.000 days that were before the coming of Christ / Messiah, we get the result: 1.440.000 + 720.000 = 2.160.000 days. This result tells us that 6.000 years has 2.160.000 days, that is to be filled in 720.000 days with 2.000 years of our New Era.

If 6.000 years have 2.160.000 days, and the perfect Creator says that perfection has 360 functions in its functioning, then there are missing days that tell us that an important event will take place at six and that behind six there is nothing to happen, like

that after the 6th day in the first week, since creation, nothing happened or was created.
The Lord God, all that he created, Hi created by Friday and on Friday on the 6th day, all creation ended. And he dedicated the seventh day and rested from all his works.
Jesus Christ is both Alpha and Omega. The end merges with the beginning, so these 6.000 years should be put together with the beginning (Alpha), where they originated, exactly on the day and hour / when the time is fulfilled for that meeting / or joining the end with a re-beginning.
But in all this information we have a deficiency, the perfection of 36 or 360 as it is most simply called has not been shown, to understand the truth more simply.
When we start looking from perfect number 36 for the location of that perfect, lost number of days to put together in Alpha and Omega, which it cannot function without the perfection in which it was created, because the perfect creator God who is in heaven and who maintains with his power and his wisdom everything we see, what we do not see and what we shall see only in the new life. We can't now, in this body, find the perfection of God, when God's perfect justification from our sins, when we go through death, because everything created on this planet because of the sin of the first man must die and is born and shows people that they will die too, and they think about not worrying too much about worldly things nor for wealth from which they will have no benefit because their lives, are limited by boundaries and their

lives pass like grass in a field. When Man dies, we quickly forget that he even existed.

Going further, we come to the discovery that if 6.000 years has 2.160.000 days (does not) fit in 36 or 3.600.000 as the number of perfection should be. The number of perfection we have been monitoring; From the beginning of life on this Earth.

When we return to the days of creation and the first week, the Creator revealed all the secrets, that after the 6th day there was no more work nor creation, but everything, by God's will, continued in the direction of which was created to function, and on the seventh day, on the Sabbath, God sanctified and on that day rested from all the deeds he did.

God dedicated the Sabbath, saying that it was holy, that no one on that day does any work but remember the Man created and to whom he gave all the blessings he enjoys, and that on that day be a day of thanksgiving to God and a day of acknowledgment of God as your creator.

This information reveals that beyond 6.000 years there will nothing, and that every work on this Earth will be finished, and begin a time in which the Earth will rest and rest from all the sins that were committed on her. The same as the first week of creation showed, and it was that first week that God revealed to us , all his secrets and all the instructions that we needed for our salvation.

“But he also said that people will have eyes and that he will not see and that he will have ears and will not hear.”

Everything is in front of us, but we do not try to hear or see. We think the most important thing around us right now is; that we are the center of the world, and that our opinion and our decisions are paramount. But our decisions and our opinion have taken us in the direction of eternal death. Our correct procedure for every decision we make is to ask if it is right before God and only then to do so as it is right, even if it was contrary to our desire. Our desires for the most part, lead us to ruin and in the opposite direction from God. All the desires of our body are only the result of the sin that lives in us. Our body always yields to the desires of the flesh, because it is of the Earth and all the desires of the flesh are only sin. But the desires we bring through conscience or spiritual rethinking, they make their way to eternal life, because this body will remain on the Earth the day of our transformation, when we are divided. Sinful body it remains on Earth because it is also part of the Earth and returns as to the Earth.

A spiritual Man, who lived according to the will of God, who did not please the body but God, he will enter into eternal life

when Jesus / Isaa comes in the clouds and raises from the grave those who have fallen asleep in the faith. He will pick them up on the day appointed, by God’s providence. He will raise up those whose sins have already been forgiven because they also prayed while they were alive, that the Creator, Heavenly Father, may forgive them all their sins through sacrifice - a lamb that is born

and gives its life voluntarily for us, that all who believe in him may not perish but have life eternal.
That day has come or is already on the threshold because the time of the (time's) of his Alpha and Omega have been fulfilled. The one who created all these Alphas and Omegas, and to guide us in that the direction for us to understand also brings time to the point where everything will stop and time will disappear.
This information says that after these 6.000 years, comes 1.000 years consecration of this Earth, in which the Earth will rest from people, and the sin that has been committed on it.
This 1.000 years, enters into the years that are counted separately from 6.000 years, just as God separated the 7th (day), the Sabbath, from 6 working days .
But in these 6.000 years, this 1.000 is added, that is.
6.000 years + 1.000 years = 7.000 years, or 2.160.000 days + 360.000 days = 2.520.000 days; but 10.000. years should have 3.600.000 days, which are 3.600.000 – 2.520.000 = 1.080.000 days; and those are the days, exactly from, 3.000 years because in 3.000 years there are 1.080.000 days.
Since in a perfect God everything is perfect, just as he himself is perfect, the path to that perfection is also revealed by the perfection of God.
These are the years Adam spent in Paradise / Garden of Eden until the day he refused obedience to God of his own free will he violated God's commandment even though he was warned, even though the Creator of heaven told him, "On the day you taste of this tree you will die that day."

This information tells to us that Adam lived in the Garden of Eden for 3.000 years and that they fit into God's perfection of 3.600.000 days in 10.000 years.

And this is this information: 3.000 years , Adam / Adem's life in the Garden of Eden, Paradise, has 1.080.000 days.

The 6.000 years in which Adam / Adem' children, to live a sinful life are 2.160.000 days.

1) Sum: 3.000 years has = 1.080. 000 days, Adam's life in the Garden of Eden.

2) 1.000 years has = 360.000 days: is the time of consecration of the Earth.

3) 6000 years has = 2.160.000 days. 6 thousand years from the beginning Adam's life (as) a sinner.

4) (3.600.000 days has = in 10.000) years.

5) (1.080.000 days has = in 3.000)year.

6) (360.000 days has = in 1.000)year.

7) (2.160.000 days has = in 6.000) year.

---

8) (3.600.000 days has = in 10.000) year.

The Lord God revealed this information to us when the time of the times has expired and when the time comes for it to show up, in its time.

Everything has its time, so the information that the Creator revealed at that time was shown through all times, because it was needed to serve in that time and the people who lived in that time. Every new information also has its time when it will be disclosed. Due to a specific reason for which it exists and for which it was sent. And this information is accompanied by another that follows and confirms it.

Everything God has created speaks 3 and 6 or 36 But it shows everything together 9. There is also a year that is of great importance.

At 6.000 years is fulfilled on 9.000 years, because those years, which Adam / Adem spent in the Garden of Eden, also-counted, only they go to those special saints, just as the holy day of the Sabbath goes.

And that is:

1.000 years of holy Sabbaths + 3000 years of holy Paradise time its,

= 4000 Holy years.

These 4.000 holy years have 1.440.000 days, and they are especially numerous.

But in the end they were put together through Alpha and Omega, with them he merged an end with a beginning.

Just like Saturday put together with 6 working days in 1the first week, of life on Earth. And when she especially separated herself from counting the days of the week, it was still counted in the days of the year.

So, on the same perfect count, all these years stand and confirm and they give the truth that no one who seeks it can go astray. It shows the years and times, together through its confirmation and confirms all this information. It contains information about the time and space they lie on of the same dimensions and on the same timeline.

They lie on the stone of Jasper or the stone of Alem. Both names are one and the same stone which is called the corner stone, which them MASONS; rejected because it bothered them in their building. Just on that Jasper / Alem stone, lies the whole building of truth created, by the Heavenly Father, and showed us through his creations. It was on that stone that everything was built, because that corner stone holds time, space and the whole Universe.

And as the truth of perfection says: the space we live in is one large 360 ° circle, when we divide that circle into parts, as it has an hour, we will get portions of space in which we live.

Each of these 12 parts of space has 30 ° or 12 x 30 = 360 °, and that fits exactly into the hour the time or either in an hour of the year or in an hour of the space. Everything makes the same information, right on that stone, corner stone Jasper / Alem, pass through it all angles from 9 to 90 ° or all angles, which may be in Alpha and Omega, and they can be unlimited, depending on their size, their space, or the time in which they are shared.

As they are shown on the clock as the corners of time - there are 12 of them or corners of the month in one year there are 12 of them or corners of space of 30 ° - there are 12 of them, they can all be divided into infinity but that is the basis of their division or their roots.

It is confirmed that the stone was rejected by people who did not believe in Him, and that on it stands all creation and that Jesus / Isaa / Messiah – is—Jasper /Alem stone, and that only he is the cornerstone. And that›s the most the more precious, hardest and most beautiful diamond that exists at all.

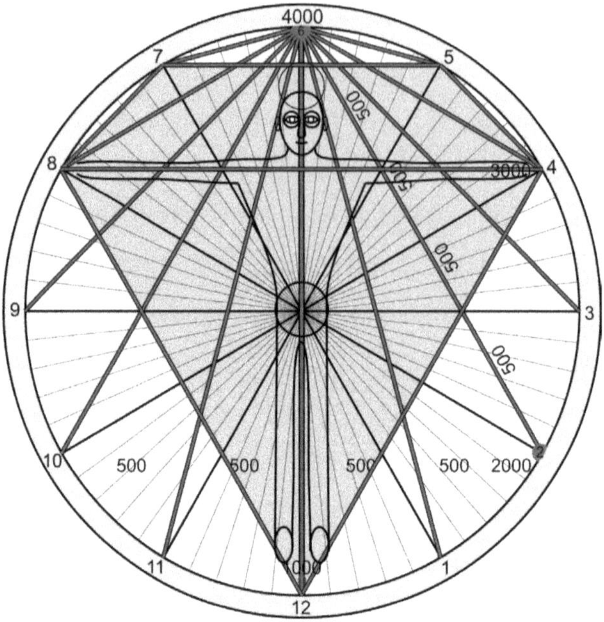

Picture 39 (I AM ALPHA AND OMEGA SAYS: JESUS CHRIST / MESSIAH). It shows the appearance of Jasper-Alem-stone and in it our Creator, our Redeemer, and our Saviour, Jesus / Isaa / Messiah.

In the book "New Testament" or Indill or in "Revelation of John" it is written:

CHAPTER 1:

Verse 1. "The revelation of Jesus Christ, which God gave unto him, to show unto his servants, things which must shortly come to pass, and hi sent and signified it by his Angel unto his servant John, who witnesses the word of God and the testimony of Jesus Christ, and of all the things that he saw. "

Blessed are those who read and those who listen to the words of the prophecy, and keep them which is written in it, because the time is near.

From John to the seven churches that are in Asia, grace and peace to you from the one who is and who was and who will come and of the seven spirits who are before his throne.
From Jesus Christ, who is a faithful witness and the begotten of the Dead, and the prince of the Kings of the Earth. Unto him that loved us, and washed us from our sins in his own blood, and hath made us kings and priests unto God and His Father; to him be glory and domination for ever and ever. Amen.
Behold, he cometh with clouds; and every eye shall see him, and they also which pierced him: and all kindreds of the Earth, shall wail because of him. Even so, Amen.
"I am the Alpha and the Omega, the beginning and the end," says the Lord God, who is and who was and who will come."
I, John, who am both your brother and friend in trouble and in the kingdom and suffering at the cross, I was on an isle called Patam, for the word of God and for the testimony of Christ. I was in the spirit and heard of a voice as great as a trumpet, saying, 'I am the Alpha and the Omega, first and last.'
And what you see write in a book, send to the churches that are in Asia, to Ephesus and to Smyrna, to Pergamum and Thyatira, to Sardis, to Philadelphia, and to Laodicea. And I looked around to see the voice that spoke to me and looked around I saw seven candlesticks of gold and in the midst of seven candlesticks as the son of man, clothed in a long robe and girded on his chest a gold belt.
Both his head and his hair were white as wool, like snow, his eyes, his feet like a flame of fire, his feet like honey as if they burned in a furnace; his voice as the roar of the water of many, seven stars

in his right hand, and out of his mouth went a sharp two-edged sword, his sides and face were as the Sun shineth in his strength.
And when I saw him, I fell at his feet as dead, and put his right hand upon me, saying unto me, Fear not; I am the first and the last.
I am he that liveth, and was dead; behold, I am alive for evermore, Amen.
And have the keys of hell and of death.
Write the things which thou hast seen, and things which are, and the things which shall be hereafter. "
CHAPTER 21:
Verse 1. "And I saw a new Heaven and a new Earth: for the first Heaven and the first Earth were passed away; and there was no more sea."
And I John saw the holy city, new Jerusalem, coming down from God out of heaven, prepared as a bride adorned for her husband.
And I heard a great voice out of heaven saying, Behold, the tabernacle of God *is* with men, and he will dwell with them, and they shall be his people, and God himself shall be with them, and be their God.
And God shall wipe away all tears from their eyes; and there shall be no more death, neither sorrow, nor crying, neither shall there be any more pain: for the former things are passed away.
And he that sat upon the throne said, Behold, I make all things new.
And he said unto me, Write: for these words are true and faithful.
And he said unto me, It is done. I am Alpha and Omega, the beginning and the end. I will give unto him that is athirst of the fountain of the water of life freely.

He that over cometh shall inherit all things; and I will be his God, and he shall be my son.
But the fearful, and unbelieving, and the abominable, and murderers, and whoremongers, and sorcerers, and idolaters, and all liars, shall have their part in the lake which burning with fire and brimstone: which is the second death.
And there came unto me one of the seven angels which had the seven vials full of the seven last plagues, and talked with me, saying, Come hither, I will shew thee the bride, the Lamb's wife.
And he carried me away in the spirit to a great and high mountain, and shewed me that great city, the holy Jerusalem, descending out of Heaven from God having the glory of God: and her light was like unto a stone most precious, even like a Jasper stone, clear as crystal.
And had a wall great and high, and had twelve gates, and at the gates twelve Angels, and names written thereon, which are the names of the twelve tribes of the children of Israel / Jacob / Jakub:
And had a wall great and high, and had twelve gates, and at the gates twelve angels, and names written thereon, which are the names of the twelve tribes of the children of Israel.
On the east three gates; on the north three gates; on the south three gates; and on the west three gates.
And the wall of the city had twelve foundations, and in them the names of the twelve apostles of the Lamb.
And the one who speaks with me he has a golden reed to measure the city, and the gates thereof, and the wall thereof.

And the city stands on the four corners, its length is the same as its width. And measured the city with the reed, twelve thousand runners. The length, width and height are equal.

And he measured the wall thereof, at 144 cubits, according to the measure of a man, that is, of the Angel.

And the building of the wall of it was of Jasper: and the city was pure gold, like unto clear glass. And the foundations of the wall of the city were garnished with all manner of precious stones.

The first foundation was Jasper; the second, Sapphire; the third, a Chalcedony; the fourth, an Emerald. The fifth, Sardonyx; the sixth, Sardius; the seventh, Chrysolyte; the eighth, Beryl; the ninth, a Topaz; the tenth, a Chrysoprasus; the eleventh, a Jacinth; the twelfth, an Amethyst.

And the twelve gates were twelve pearls; every several gate was of one pearl: and the street of the city was pure gold, as it were transparent glass.

And I saw no temple therein: for the Lord God Almighty and the Lamb are the temple of it.

And the city had no need of the sun, neither of the moon, to shine in it: for the glory of God did lighten it, and the Lamb is the light thereof.

And the nations of them which are saved shall walk in the light of it: and the kings of the Earth do bring their glory and honour into it.

And they shall bring the glory and honour of the nations into it.

And there shall in no wise enter into it anything that defileth, neither whatsoever worketh abomination, or maketh a lie: but they which are written in the Lamb's book of life.

And he showed me a pure river of water of life, clear as crystal, proceeding out of the throne of God and of the Lamb. In the midst of the street of it, and on either side of the river, was there the tree of life, which bare twelve manner of fruits, and yielded her fruit every month: and the leaves of the tree were for the healing of the nations.
And there shall be no more curse: but the throne of God and of the Lamb shall be in it; and his servants shall serve him:
And they shall see his face; and his name shall be in their foreheads. And there shall be no night there; and they need no candle, neither light of the Sun; for the Lord God giveth them light: and they shall reign for ever and ever.
And he said unto me, These sayings are faithful and true: and the Lord God of the holy prophets sent his angel to shew unto his servants the things which must shortly be done.
Behold, I come quickly: blessed is he that keepeth the sayings of the prophecy of this book.
And I John saw these things, and heard them. And when I had heard and seen, I fell down to worship before the feet of the Angel which shewed me these things.
Then saith he unto me, See thou do it not: for I am thy fellow servant, and of thy brethren the prophets, and of them which keep the sayings of this book: worship God.
And he saith unto me, Seal not the sayings of the prophecy of this book: for the time is at hand.
He that is unjust, let him be unjust still: and he which is filthy, let him be filthy still: and he that is righteous, let him be righteous still: and he that is holy, let him be holy still.

And, behold, I come quickly; and my reward is with me, to give every Man according as his work shall be.
I am Alpha and Omega, the beginning and the end, the first and the last.
Blessed are they that do his commandments, that they may have right to the tree of life, and may enter in through the gates into the city. For without are dogs, and sorcerers, and whoremongers, and murderers, and idolaters, and whosoever loveth and maketh a lie.
I Jesus have sent mine Angel to testify unto you these things in the churches. I am the root and the offspring of David, and the bright and morning star.
And the Spirit and the bride say, Come. And let him that heareth say, Come.
And let him that is athirst come. And whosoever will, let him take the water of life freely.
For I testify unto every man that heareth the words of the prophecy of this book, If any man shall add unto these things, God shall add unto him the plagues that are written in this book. And if any man shall take away from the words of the book of this prophecy, God shall take away his part out of the book of life, and out of the holy city, and from the things which are written in this book. He which testifieth these things saith, Surely I come quickly. Amen. Even so, come, Lord Jesus.
The grace of our Lord Jesus Christ be with you all. Amen. {THE END.}

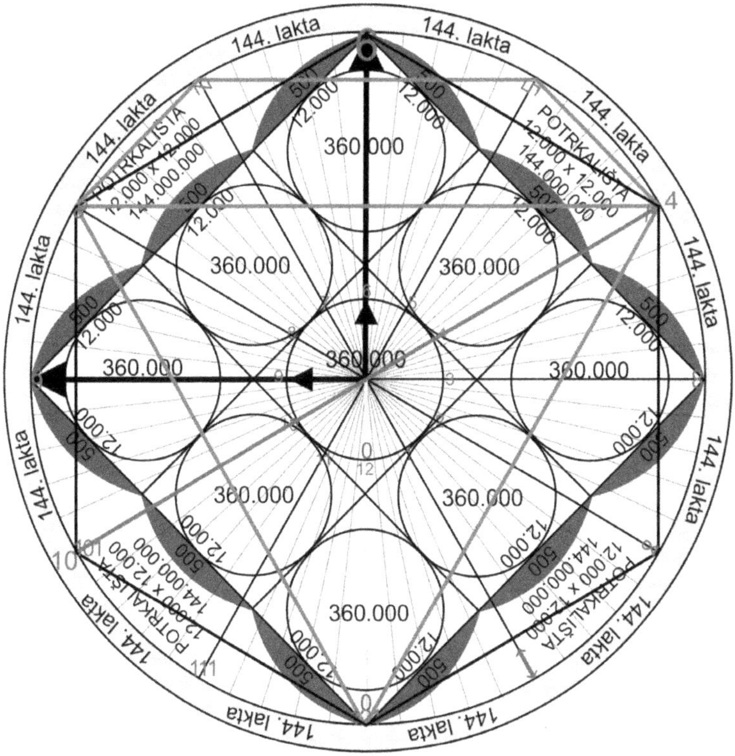

Picture 40 (TIME AND SPACE ON STONE JASPIS)

This picture shows how the word of God describes the new city Jerusalem that will come down from heaven and will be brought down to Earth.

Alone the word Jerusalem and the name given to it by the Lord God the same who gave his name to Jacob / Jakub and said to him, (from this day is your name Israel). The same as Saraji was given the name Sara, and Jerusalem was formerly called Jebuzaj. On that day when the sons of Jacob's / Jacub's / Israelites, came from Egypt to a land led by the Lord God, which he swore to Abraham / Ibrahim that he would inherit it . To him and his sons, he fulfilled that promise and divided the land through his faithful servant Joshua, son of Nun.

Upon entering the Promised Land, into which the 12 tribes of Israel entered, Jebuzaj was named Jerusalem. In that city, chosen by God, in which the first building was built in which the name of God began to be invoked, in which was entered the law of God which the Lord God wrote with his finger, on Mount Sinai, 10 Commandments of God. The law which is written by the hand of God is holy, and it is in the temple of God, in Jerusalem, because God himself ordained the place for it. And the spirit of God descended in that temple, in a separate place where only high priests were allowed to enter, and that once in year.

In front of this temple, in Jerusalem, there were slaughtered sacrifices, every day in the evening, before on the coming of the new day; and which The Lord God burned with his flame. In that incineration there was a sign that the victim- sacrifice had been received. Temple in Jerusalem was dedicated to God's presence, which every day confirmed in the fire that was coming down from Heaven and burned the sacrifices. The sacrifices for sins or sacrifices for thanksgiving.

Jerusalem was a special city in which the greatest events on this Earth were to take place where they were supposed to happen the biggest events on this Earth of our entire universe. In Jerusalem, the plan of salvation was to be fulfilled for, this, Adam's / Adem's generation. It was fulfilled after Abraham's / Ibrahim's grandchildren, the sons of Jacob; Judah, Benjamin and the half tribe, of Manasseh came to Jerusalem.

These two and half tribes, moved to this part of the Promised Land which belonged to them and from these two tribes and half, were born the kings who lived on that part of the Promised Land. From the tribe of Benjamin was also the first king Saul.

And of the tribe of Judah was born the emperor David / Daud, who wrote the Psalms and Prophecies of Zebur and which many, years ago, before Jesus / Isaa birth, wrote about his arrival and his death, and confirmed that Jesus will overcome death in his death and that he will in his death; Lock and Unlock two doors at the same time: It will lock eternal death and unlock eternal life. And that in his hand are the keys to life and death, that he created us and that he will raise us from the dead when the time came. And that Jesus Christ is the Messiah and that he is the first and the last, and both; beginning and end, Alpha and Omega, of all creation.

All those who are written in the book of life, all will rise when Omega with the Alpha is merged of our time or the time of this planet.

After this time, when the holy time comes and when everything created, which the Lord God created in 6 days, as it is written In the book of "Genesis" or "1. A book of Moses "or" Torah " (These are the names for one same book).

As the days are created, so one day they will pass, as well as everything that passes. And there will come another time, a holy time in which there will be no sin or sinner or originator of sin - Satan or the originator of lies who deceived our father Adam / Adem and our mother Eve / Hava.

It says that he who gave his life in Jerusalem, on Golgotha delivered us from eternal death, at a specific time, year, day and hour. But since the planets have done what should have been done and since time passes through all that is created, and the time comes for all that is created, which was and is created for God's specific functions, and that all functions are over, time is over. Time also ends, for time also had its task and function; for which it was created.

Time exists only in this galaxy of ours and time is not limited to the Lord God nor the sinless worlds that are still in outer space. God's word says that life on this planet will be restored in which keeps showing us the nature, plants, and animals. Everything visible and invisible has the same seal of God, the same direction, life, death and life. This is where the first Alpha began, because the letter A is also the first letter in the alphabet, which means first.

Then death and the end or Omega. By definition: everything created has its end when the time passes for which it was created or when the time of a specific task passes and nothing is created which we see or do not see without being created without exactly its certain functions.

Everything created is consumed in its function when it performs one certain function, for which it was created and for which precisely determined

the time for that function to be performed, in a precisely determined time.

Alpha means:

A - first:

L - last – end :

F - frequency repetition:

A - initial creation:

That is, everything created crosses the boundaries of its duration and comes to an end, and disappears.

As it is assembled with the beginning, where Omega touches the Alpha, the end of a creation that already had its time is formed and a new beginning of the same begins a creation but in a new time.

This happens constantly and frequently with all of God's creations and people are God's creations too. Only that the Creator created everything with his own word that was: BE and it WAS. And the Lord God saw that all was well what he created. And all created, created to serve Man. And since Man sins, The Creator, whose name is love, out of love for sinners he also gave Man his life to redeem him from eternal death.

So he created a plan of salvation for a sinful Man, and that plan was fulfilled in Jerusalem, on the cross, on the hill of Golgotha or Ossuary .

Picture 41 (THE END OF LIFE THROUGH ALPHA AND OMEGA).

In the beginning God created (Heaven and Earth), and the Earth was desolate and without form, and darkness was over the abyss. Every Man born on this Earth dies. But he just falls asleep, as nature shows that in every year there is a period when all nature sleeps - in winter. In those months, nature looks dead but is in fact, asleep. So people, when their life functions cease, they fall asleep. But at a certain time, when the time comes for that, all will be awakened, some to eternal life, some to eternal death.

This is what God says through all creation that confirms itself. That means that after sleep or death we will wake up again just like nature waking from hibernation. Why is it called hibernation?

Because, when the time of time on this Earth passes, our entire planet Earth will have a period of winter, when there will be no life at all on it, and that will last for 1.000 years.
In that period there will be no natural conditions for life, the Sun will lose its light, because light is both the first creation and the initial element of life on this Earth. Sun, heat and water are the main elements that sustain the life of all creation.
When the one who created light removes light, at that moment all life functions of life ceases and darkness appears, which also was before the creation of life.
And if there is no Sun or heat then it is polar winter. All life stops and stands in ice for 1.000 years as shown by the whole universe, i.e. time, shows that time has its time and it will come to his Omega, which it followed and to which it will come, exactly to the place of its creation when it orbits its 360 ° circle.
But time also has its sleep time, as does nature, right on Its time to wake up, so will people in their time awaken, after 1.000 years. But those who will sleep in ice 1.000 years, they will awaken to eternal death. Nature itself says that nothing can be 1.000 years in the ice and then have eternal life, i.e. continue to live in the Garden of Eden.
These people will wake up but they are all for eternal death, no one will inherit Paradise or Jenet because eternal death awaits them, in a certain time. The death he will have after SLEEP, there are now people who are dying and who have been dying for these 6.000 years, and it's called the second death. This means:

the first death is sleep, awakening and then judging according to merit; by the law of God.

The Ten Commandments of God is the measure by which people will be judged on the day of judgment, by their deeds. The law of God given to Moses / Musa on Mount Sinai upon leaving Egyptian slavery, is the foreshadowing of the law of God found in heaven and on it stands the throne of God, which is also the law of eternity.

It is written in the word of God, "The court sits, and the books are opened."

Justifications through Jesus / Isaa / Messiah, and all those who before their own death justified from their sins, through the blood of Christ and through Christ merit, they are on the second coming of Christ, with Jesus in whom they believed that he could give them life again and through faith in Christ, that he created them and that he can create them again and bring them back to life. They got life because Jesus / Isaa / Messiah is the creator of life, in his hands is the life of everything created visible and invisible. And everything that is invisible and it does its function to make it visible to function.

All who have their sins even while they were alive confessed before God and sought forgiveness; through the holy blood of the Lamb slaughtered on Golgotha and through his blood were justified; will not pass another death after awakening, when Jesus /Isaa / Messiah, comes on the clouds, in the last moment of life on this Earth or on a day called Judgment Day / Kijamet

Day. When he awakens all whose names are written in the book of Lamb or in the book of eternal life, only they receive eternal life and with the Creator, Redeemer, and Saviour go to the clouds and into eternal life.

That group of people, whose sins are forgiven, while they are still alive on this Earth they will be in heaven for 1.000 years, in holy form which they received immediately after being awakened from sleep by their Saviour they had been waiting for. Some were in mortality for a very short period in sleep and some slept for a long time. No matter how long someone slept, when he wakes up, he will not feel the length of his sleep, just as we have no sense of how long we have slept. Every night in our sleep we lose track of the time and length of time we spent sleeping.

It is a foreshadowing of that dream, in which we will sleep before, we awaken, divided each in his own time, one for eternal life and other for eternal death. After this period, when the exact time for that passes and when the time comes to awaken those who are for trial and who for their deeds they must answer before the Lord God, the creator of all things visible and invisible, who suffered by his grace and gave instructions for salvation.

And gave his firstborn son, whom he first created in Heaven, for the firstborn son he created on Earth and for his sinful children, who gave all their love in suffering so that no one dies but through the sacrifice of Christ / Jesus / Isaa, receive forgiveness of sins. Because forgiven sins are not sins. All those who loved to live according to their desires and according to their free will, which God has given to all of us, to make a decision of our own will and behaviour by our own free will, and who have chosen by their own free will to die an eternal death, they will end their time of life which they have ordained in this life and in this time through which they have passed on this Earth.

After all, God says, “I am creating a new Heaven and a new Earth, and the sea will be no more, I am lowering the new city of Jerusalem from Heaven, to the same place where it is now.” In the same place because in that place was the Garden of Eden and there it will be again a new life, in the form in which it was originally created, i.e. eternal and sinless life. Alpha and Omega goes through space, mass, genes, size and creates everything new. Only after the time that was limited and that came back to its starting point, a point that starts from scratch and goes in his perfect direction and comes to a starting point. It made a 360 ° circle along the way. This Alpha and Omega went around a very big circle, that’s why its name is Omega. This Omega was very big for in it all things were created and made up all the elements of life, of all creations that are in perfect functioning where the creation of the life of all living things are on this Earth. This means (O) -omega or large volume, a large surface in which the beginning was of all beginnings and in which is also the end of that beginning, in such a form. Everything created existed within its limit set by the Lord God who created that existence.

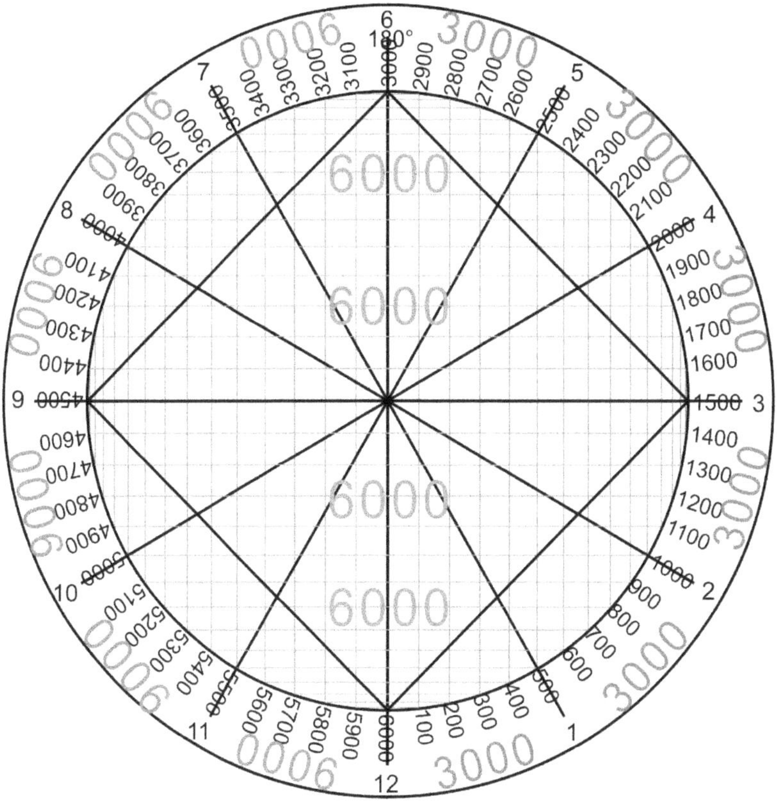

Picture 42 (LIFE THROUGH TIME AND SPACE).

This picture shows the time, which has been from the very beginning of the creation of time.

From the first Second: through the Second, Minute, Hour, Day, Year, and through Thousands of Years.

Then, we see a time that goes from zero (0) to 100, 200, 300, years, where it comes first at 3.000 years, in the vertical direction; And where all numbers add up and show 3.000 years and that addition ends in the middle of an hour of time.

And this direction of time is also the direction it shows the time that goes towards us, i.e. that it is the direction of life, at the moment or future tense. The past tense is going in the opposite direction from us and from future life and eternal life.

Then, through the hour of time we see numbers showing 6.000 years. And that all the hours are collected, which fit into each other in the horizontal direction, creating a sum of 6.000 years. All numbers that fit in the vertical direction, from the middle to the right, show the sum of 9.000 years. Looking from the beginning of time, we see:

A second is a creation of all times and 1 minute has 60 seconds:

6 minutes x 60 seconds = 360 seconds:

Or 60 minutes x 6 hours = 360 minutes:

Or 1 year is 12 months x 30 days = 360 days:

Or 10 years x 360 days = 3.600 days:

Or 100 years x 360 days = 36.000 days:

Or 1.000 years x 360 days = 360.000 days:

From this information, hours of time, we learn that all hours are hours time through seconds, minutes, hours, days, years, made up one and the same whole or beginning and end, as Alpha and Omega in their own perfection of 360 ° degrees or seconds, as the unit by which the beginning of the calculation of time, as well as all time.

Because 360 is a number called both the beginning and the end, and in this picture is shown in its perfection, because the vertical directions, which depart on the left side of time and fit 100% into their whole which goes in both directions up or down, those two directions make up a whole of 3.000 years. An hour that shows the direction along a horizontal line in both direction of time, shows numbers of 6.000 years, which fit into 3.000 years.

And these two directions, vertical and horizontal, together form a whole of 9.000 years: 3.000 + 6.000 = 9.000.

Just like 3 + 6 makes = 9. All of the above information show the number 9 because zero has no value, it only serves to confirm the information on how much time has passed since it should show that information, serve as a seal, confirm that this information is correct and makes it easier for people to see the time, the space or number they want to reach.

This picture of time sends us very important information and confirms that the time of year on this planet is counted in the same perfection of Alpha and Omega and that the final sum is 9.000 years. And the number, 9 is indicated by all the numbers and all the sums that exist throughout history of this planet and its time, existence and life on it, as well as the time that will exist, until the time when time will lose its significance, i.e. until the very moment when the weather will return frequently to its original point of origin and merge where it was created, to the beginning, to the Alpha and to the beginning where there is no time, at the time where Adam / Adem was created, where he lived before sin, in the Garden of Eden itself, Paradise, Jenet or into eternal life.

It's on the hour line, at a point where 3.000 + 6.000 years composed into 9.000 years, and at the very zero where the beginning as well as the end of the time given to this planet until

the second coming of Christ / Messiah and the end of time which was to be used for our future time.
This picture, no. 42, shows the time that has ended its function at 9.000 years and which was created by adding 3.000 and 6.,000 years. Since it is counted from Adam's / Adem's fall into sin, and that through the time that was written down and confirmed through counting as the Lord God gave to count the time on this the Earth, through the Sun and the Moon or Day and Night, to show humans that they have their time.
When that time is added up so far, it is about 6.000 years: Or 4.000 years, Old Era and about 2.000 years New Era. All this information shows that there was a time that was not recorded in the records of this time, which is called the time of Garden of Eden.
And that, through all these 6.000 years, only sin has been committed on this Earth, from the first Man to sin, Men became sinners and committed only sin.
But the 9.000-year period, which is the end has a secret which was revealed at the time when Adam / Adem lived in the Garden of Eden, and that is: 3 + 6 = 9. Among all these perfect entities of God in the Universe, which show their accuracy through the numbers 3, 6 and 9, there is also a letter that connects them, and that is the letter (AND) that is between each number 3 (AND) 6.
There is a letter between the numbers 3 (AND) 6 and in the pronunciation itself, the number 36 or 360, we always hear the letter (AND) that separates them in their meaning. And this means that 3 and 6 are one whole, connected in time which is in the whole

of omega or Alpha and Omega, but not in whole with the kind of time. It is one of 6.000 years, it is the time of sin and its number is 6; and the second is 3.000 years it is a holy time in which there was no sin, the time of Paradise or the Garden of Eden, and its number is 3.

That means, here this information met and confirmed one second, explained why the whole Universe shows the number 9 in various ways: through numbers, through time, and through space, and that all the roots of the numbers end in number 9, such as: 0, 1, 2, 3, 4, 5, 6, 7, 8, 9. After number 9 all numbers just repeat. When we add these numbers: 1 + 2 + 3 + 4 + 5 +6 + 7 + 8 = 36; 3 + 6 = 9 or 36 + 9 = 45 or 4 + 5 = 9.

Then, when we double the roots of the numbers: 2 + 4 + 6 + 8+ 10 + 12 + 14 + 16 + 18 = 90;

or when we multiply numbers:

9 x 1 = 9; 9 x 2 = 18; 9 x 3 = 27; 9 x 4 = 36; 9 x 5 = 45; 9 x 6 = 36; 9 x 7 = 63; 9 x 8 = 72; 9 x 9 = 81;

9 x 10 = 90; All these numbers when multiplied, their sum is = 9, and number 144 = is, 1 + 4 + 4 = 9; like all the other numbers they have significance related to perfection, space, time and number, everything is shown by the number 9.

At number 9 ends the history of this planet, this galaxy of ours and the forms of this life on it. Which will be renewed in the same genes, in the same space and size, in its time when the time comes for that. God showed us everything and then confirmed that everything was shown exactly, so that there would be no doubt in the truth of what all this is about.

Each child conceived and born spent 9 months in the mother's womb. Here we see number 9, because Jesus was born, also, after 9 months spent in the mother's womb. And that exhaled at 9 ninth hour (day). The Lord God will restore the original life, greatness and appearance as it is was and was created in the original form of matter, size, genes, in which it originated and which will only continue its life function in which it was stopped in his existence.

But everything that God has created is also sacred and can never disappear, so is the life in which Man was created and for which he was created in to its original state, it will be restored like all Alpha and Omega which was renewed before the onset of sin. It will just disappear the originator of sin, the originator of lies, the enemy of God will disappear and the enemy of Man or Satan, the old serpent who deceives one third of the angels of heaven and the first created people on this Earth. Through their fall created the sin, through which people pass on their way to return to the Garden of Eden.

Time:

Shows in the space of these 9.000 years that there will be another 1.000 years that will occur after these: 6.000 + 3.000 = 9.000 + 1.000 years in which the Earth will rest and in which the Lord will sanctify the Earth, just as it is after the 6th day, after creating life on it, dedicated the 7th day Saturday, so will dedicate the seventh thousand year in which the Earth will rest and there will not be no work on it. In which the saved who will already be in Heaven glorify the Lord God who brings them back to eternal life, just like how it should have been celebrated and praised by the people of this world who, also were created in the Garden of Eden.

The Garden of Eden, which God created and also sheltered, and will return again when the time comes. This information now complements itself that perfection was put together in the very beginning as:

6.000 + 1.000 + 3.000 = 10.000 years. 10.000 years is also a whole perfection, because in 10.000 years has exactly 3.600.000 days. This is where all the Alphas and Omega of everything created as well as everything that will be re-created on this Earth. And this is confirmed by the information contained in the last book in the "New Testament" or "Revelation of John" which describes the events that will take place after the end of life on this Earth, in this form in which we now live.

In the book called "Inđil" and which is an integral part of the book "New covenant "in" Revelation of John "reads:

CHAPTER 20;

Verse 1. "Then I saw an Angel coming down from heaven, having the key to the bottomless pit and a great chain in his hand."

He laid hold of the dragon, that serpent of old, who is the Devil and Satan, and bound him for a thousand years and he cast him into the bottomless pit, and shut him up, and set a seal on him, so that he should deceive the nations no more until the thousand years were finished. But after these things he must be released for a little while.

And I saw the throne, and they sat upon it, and judgment was given to them, and souls cut for the testimony of Jesus and for the word of God. Who did not worship the beast and his icon and did not receive the mark on his forehead and on his hand and came to live and reigned with Christ / Messiah 1.000 years. And the rest of the dead did not come to life until 1.000 were finished. This is the first resurrection.

Blessed and holy is he who has a part in the first resurrection, over him the second death has no dominion, but will be priests to God and Christ / Messiah and will reign with him for 1.000 years. And when the 1.000 years are over, Satan will be released from his dungeon and he will go out to deceive the nations of the four corners of the Earth, Gog and Magog. To gather them to battle, the number of which is like the sand of the sea. And they went out into the midst of the Earth, and compassed the saints about, and the beloved city, and fire came down from God out of heaven, and devoured them.

Verse 10. "The devil, who deceived them, was cast into the lake of fire and brimstone where the beast and the false prophet are. And they will be tormented day and night forever and ever."

Verse 11. "Then I saw a great white throne and Him who sat on it, from whose face the Earth and the Heaven fled away. And there was found no place for them."

Verse 12. "And I saw dead and small and great standing before God and the books opened. And the second book is opened, the book of life, and the dead were judged according to their works."

Verse 13. "And the sea gave up the dead who were in it, and Death and hell delivered up the dead who were in them. And they were judged, each one according to his works."

Verse 14. "Both death and hell were cast into the lake of fire. And this is it the second death."

Verse 15. "And anyone not written in the book of life would be cast into the lake of fire."

CHAPTER 21:

Verse 1. "Now I saw a new Heaven and a new Earth, for the first Heaven and the first Earth had passed away. Also there was no more sea."

Then I, John, saw the holy city, New Jerusalem, coming down out of Heaven from God, prepared as a bride adorned for her husband. And I heard a loud voice from Heaven saying, "Behold, the tabernacle of God *is* with men, and he will dwell with them, and they shall be his people. God himself will be with them and be their God. And God will wipe away every tear from their eyes; there shall be no more death, nor sorrow, nor crying. There shall be no more pain, for the former things have passed away."

Verse 5. "And he that sat on the throne said, Behold, I make all things new. And he said to me, Write, for these words are true and faithful."

Verse 6. "And he said unto me, It is finished. I'm Alpha and Omega, the beginning and ending. I will give of the fountain of the water of life freely to him who thirsts."

Verse 7. "He who overcomes shall inherit all things, and I will be his God and he shall be My son."

Verse 8. “But the cowardly, unbelieving, abominable, murderers, sexually immoral, sorcerers, idolaters, and all liars shall have their part in the lake which burns with fire and brimstone, which is the second death.”

From all this information, which together compiled this great Alpha and Omega, from the complete origin of life on this Earth, to the last information that leads us to life after this life, and which fit into one great whole that makes up the life of this Universe, and which gives us complete perfection, from the beginning to the end of life on this Earth. But in this overall information, which God has revealed to us, through his word and through his confirmation, we came across information that says:

All those who did not seek forgiveness of sins while they were still alive on this Earth, and whose sins have not been removed out of the books into which are our daily reports about our behavior, and the mistakes made every day. Someone had to justify those sins, so we could be justified by them it could only happen in one way that God approved when his (firstborn son)offered his life in Heaven as a substitute for the life of the firstborn. A son of God born on Earth.

And with his life and his own love reconciled the sinner with God. All of them, who did not seek the forgiveness of sins while they were still in life on this earth, and whose sins were not erased from the books into which they were daily reports on our behaviour, we also sinned, and those sins had to be justified, yes we could justify them, it could only happen in one way that God approved when his firstborn son in Heaven offered his life as a substitute for the life of

the firstborn, the son of God who was born on earth. With his life and his love he reconciled the sinner with God.
He opened the way to eternal life through the Cross on Golgotha and Christ / Messiah's body.
We already had information, when God created us, that we are created in his own image and that he created us that we live forever, and that our brain as well as our body has the form of God. But because of the living conditions in which our ancestor Adam / Adem began to live after leaving the Garden of Eden and when he lost the food found on the tree of life; the day he stopped using that food, from that day his life began to shorten and he began to die gradually. After all this information, which has so far pointed to the truth and to the way that leads back to the Garden of Eden and that led through these 6.000 years that our planet Earth has gone through, as well as all life on it, we have seen only the perfection of God's creation as well as perfection of fitting, perfect in perfect. These 6.000 years have shown only one direction, the path of truth and life, which has become, which will disappear and which will become again. We, with the free that we have, we can decide which way to go, and that God who speaks for himself that he is the Alpha and the Omega or that he is the first and the last and that in that same direction he created everything to function as the Alpha and the Omega, i.e. that everything he has created works perfectly, just as he himself is perfect in its circle of Alpha and Omega.
Thus, after all, when time merges with Alpha and Omega, when the end touches the beginning, at that point of connection, a

new form of life on this Earth emerges. And this new form of life is, in fact, the life that Adam / Adem lost due to disobedience to God's law.

That is, in that touch, the life that was and was created will be renewed in the sinless form or life that was in the Garden of Eden.

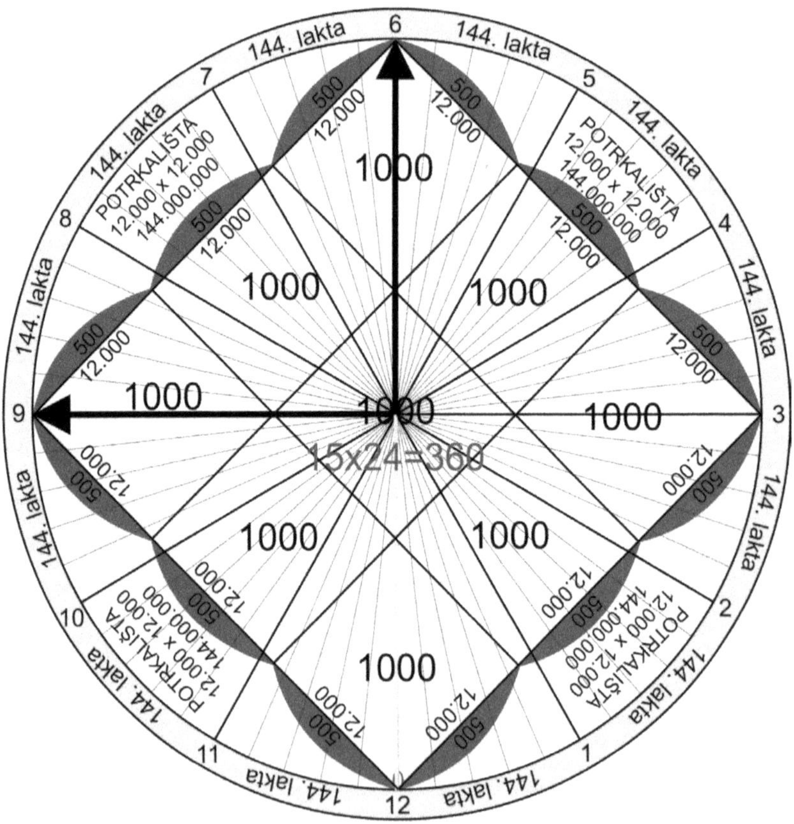

Picture 43 (NEW START).

From this picture you can see (the shape of) the new city of Jerusalem: which corresponds with the measures that are written on the last page of the book (New Testament or Indil).

This new form is described in the book "Revelation of John"
CHAPTER 1:
Verse 1. "And I saw a new heaven and a new earth, for the first heaven and the first the earth had passed away. Also there was no more sea."
Verse 2. "I, too, John, saw the new holy city of Jerusalem, where it came down from God, from heaven."
Verse 4. "And God will wipe away every tear from their eyes; there shall be no more death, nor sorrow, nor crying. There shall be no more pain, for the former things have passed away."
Verse 5. "Then He who sat on the throne said, "Behold, I make all things new." And He said to me, "Write, for these words are true and faithful."
And He said to me, "It is done! I am the Alpha and the Omega, the Beginning and the End. I will give of the fountain of the water of life freely to him who thirsts."
Verse 7. "He who overcomes shall inherit all things, and I will be his God and he shall be My son."
Here, from these verses we have information that the new city of Jerusalem will come down from Heaven again, as we read when Adam left the Garden of Eden because of his sin. God removed the Garden of Eden from the Earth and will restore it, just as all the Alphas and Omegas were restored , so the Garden of Eden will be restored and returned to the place where it was when it was created.
All this information that the Universe has shown us shows that the place was where God appeared, in the Temple, in Jerusalem, a

place called the holy of holies and which only (the high priest) was allowed to enter, once a year, a day the cleansing of the people from sin and in which Jesus / Isaa personally, by being present in it, he showed that the place had been consecrated from the very moment of the creation of life on Earth. And that it will, that especially consecrated place, be consecrated again, and that the new holy city of Jerusalem will descend to the same place, because only there, through the Alpha and Omega, can descend.

Verse 11. "And he had the glory of God. Her light was like a most precious stone, like a Jasper stone, clear as crystal. And had a wall great and high, and had twelve gates."

Verse 14. "And the wall of the city had twelve foundations, and in them the names of the twelve apostles of the Lamb."

Verse 16. "And the city lieth foursquare, and the length is as large as the breadth: and he measured the city with the reed, twelve thousand furlongs. The length and the breadth and the height of it are equal."

Verse 17. "Then he measured its wall: one hundred and forty-four cubits, according to the measure of a man, that is, of an Angel."

Verse 18. "The construction of its wall was of jasper; and the city was pure gold, like clear glass."

Verse 23. "The city had no need of the Sun or of the Moon to shine in it, for the glory of God illuminated it. The Lamb is its light."

Verse 27. "But there shall by no means enter it anything that defiles, or causes an abomination or a lie, but only those who are written in the Lamb's Book of Life."

CHAPTER 22:

Verse 1. "And he showed me a pure river of water of life, clear as crystal, proceeding out of the throne of God and of the lamb."

Verse 2. In the midst of the street of it, and on either side of the river, was there the tree of life, which bare twelve manner of fruits, and yielded her fruit every month: and the leaves of the tree were for the healing of the nations.

The city of Jerusalem will descend from heaven at the right time, at the moment when the time has passed which was the appointed time i.e. the existence of life on this Earth as well as in this galaxy. When the end with the beginning or Omega with Alpha merges, at that moment will be the time that this planet Earth had, as well as this galaxy of ours in which this whole life or the life of all creation took place, which was created in the first week of creation, it will end, just as all lives were born and changed lives, everything created that can be seen and not seen with the naked eye, and it existed so that this world could function fully and perfectly.

So at the moment when the beginning and the end merge and when the circle of life on this Earth ends, which is also a 360 ° circle, the Sun will perform its task, just like all other creatures that had their God-ordained task for which they were created. And that shows the first moment of the creation of life on this Earth where it says:

Musaa / Moses Book; CHAPTER 1:

Verse 1. "In the beginning God created Heaven and Earth. And the Earth was without form or shape and desolate. And there was darkness over the abyss."

From these verses which darkness was on Earth before God created life on it, it shows that there will be darkness on Earth again, when it ends, which it should end, in its Alpha and Omega. The moment Omega touches Alpha, in its circle that it circled and made 360 ° and where it touched the starting point where it ends life on this Earth, in this form.

But even though it will be darkness, as it was darkness before light, that darkness will be limited in time and space, i.e. there will be darkness only in our galaxy, because only in our galaxy did sin happen. Since everything has a definite time of its existence, so that darkness will last for exactly 1.000 years. And why 1.000 years?

It is the sanctification of the Earth or the rest it has not had since God said, "Six days work and the seventh day is the Sabbath, do not do any work that day", it is a sign that stands between God and people who want to obey our God's orders.

Since the Sabbath day was not celebrated, this seventh thousand years will be celebrated and our entire galaxy will rest. The same as God left in the first week, when he created everything in 6 days and he sanctified the seventh day, and on that day he did nothing but left it to the people to rest, and to himself as a memorial, as a sign of confirmation that the Lord God created the life of all creation. And that in his hands and in his grace all things were created, that he alone created, revives and gives a deadline to all creation.

1.000 years has exactly 360.000 holy days; or

(1 week has 24 holy hours);
10 weeks has 240 holy hours;
100 Weeks has 2.400 holy hours;
1.000 weeks has 24.000 holy hours;
10.000 weeks have 240.000 holy hours;
10.000 Weeks Have 10.000 holy Days.
10.000 days divided into 360 days, how long one year has:
10.000: 360 = 27.777.777 x 6 = 166.666.667 x 6 = 1000.
This information showed that, exactly, in these 6.000 years that have passed and 1.000 holy years that have not been celebrated, and that God, who left it to be celebrated, after a time when there will be no life, there will the holy days just like the hours in that seventh thousands years be fulfilled. Just like God said. And that every word of God, which he said and which had its specific purpose will be fulfilled.
So the Universe and God say that everything created must fit in one perfect whole to confirm itself, for God does not need a witness, everything created is a testimony of his perfection.
The information says: in the beginning and through the beginning, i.e. through Alpha (read the universe) confirm the truth; all that has been will be again. So we go back to the first week of creation, as well as to the day Friday in which both the first man and the first woman were created.
And that would be on the 6th day, at 144 hours, on Friday.
This is written on the first page of the book of Moses / Musaa
CHAPTER 1:
Verse 27. “So God created Man in his own image, in the image of

God created he him; male and female created he them."

Verse 31. "And God saw everything that he had made, and, behold, *it was* very good. And the evening and the morning were the sixth day."

Since then, the word of God has been written as well as the history of this planet.

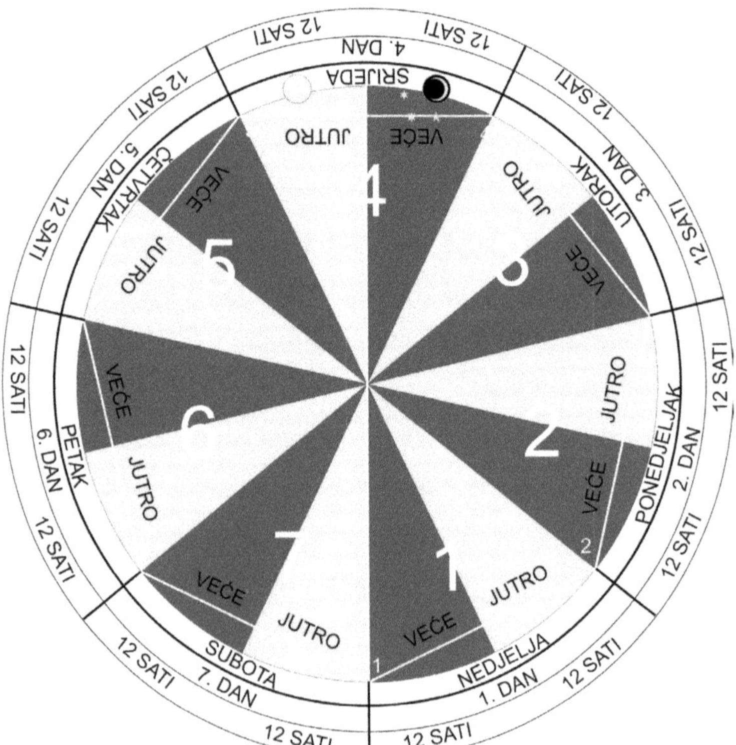

Picture 44 (FIRST WEEK OF CREATION).

On the last page of the book:

NEDELJA–SANDAY; PONEDELJAK–MONDAY; UTORAK–TUESDAY; SRIJEDA – WEDNESDAY; CETVRTAK – THURSDAY; PETAK – FRIDAY;

SUBOTA - SATURDAY

"Revelation of John" or " The New covenant ", or Inđil, which describes the end of all creation, writes:

"And the city stood on the four corners, and its length was as great as its breadth, and it measured the city with a reed on 12.000 races length, width and height were equal. And it measured the wall thereof, 144 cubits, according to his measure of man which is also angels."

From this information, we came to know that after time in which we enter again it will come again a time that fits into the time that has already passed and which, at a certain moment re-emerges, i.e. we come to the end of time and at the same moment merges Omega with Alpha and it will be the end and the beginning of time again.

Also, we see the end of everything created that takes place exactly, a certain trajectory of life on this planet Earth, and end of the whole of our milky galaxy.

Although at that moment, life, which took place in this galaxy of ours, will be stopped for a while, that is a 1.000 years, everything in this galaxy will rest on a principle that has existed before creation of life in it. He who created life will create it again, when the time comes for it.

Exactly at a certain time, which will be given at that time, exactly for a specific purpose, i.e. the time of judgment. It is also a time of God in which we have no access, as we did not have access at the time in which God created light, water, nature, day and night and when on the fourth day, Wednesday, he created the Sun, Moon and all the host of Heaven.

The time, from Wednesday, is also our time after which God created life on Earth, how people so animals, as well as every creature that breathes. But on the 6th day, Friday, created Man and Woman in the 144th hour: since the beginning of life on this Earth.

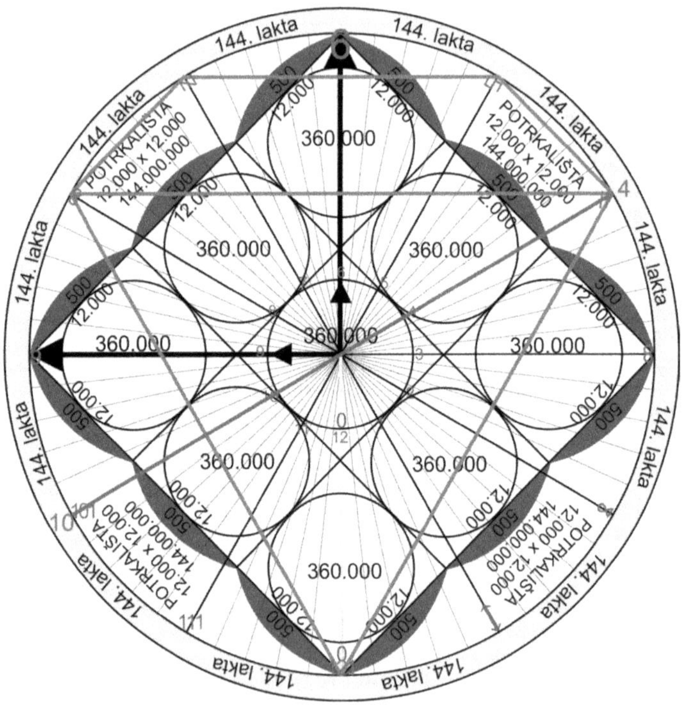

Picture 45 (SPACE, TIME AND PEOPLE ON THE SAME LINE).

From this information, which speaks for itself and explains itself, we see that the Garden of Eden, Paradise or Janet or the new city of Jerusalem will have the shape of a cube and that the number 144 has been transferred from this world to a new world, i.e.

into eternal life. Also, the number 12 shows information for itself: And it would be evening and it would be morning, i.e. shows the time that separated: night have 12 hours and day have 12 hours. During these first 12 hours, which it separated Night and Day, time counted all these 6.000 years and will be counted until the beginning of time, i.e. until the time of the times passes.

But the 12.000 x 12.000, which describes the word of God on the back of the book of Revelation or the New Testament, is also more than words, where the word of God says it is through time, and on time completed the history of this planet and life on it but that life will continue exactly according to the time counted on this Earth, as that Men who are (in the image of God) and (in the measure of Man which is also the Angels) will enter the door to this new life, the door that Christ / Messiah opens, because in his hand are all the keys, key of eternal life or eternal death. Here we also get new information that says the new city of Jerusalem looks like a cube.

The Book of Revelation

CHAPTER 21:

Verse 16. "And the city lieth foursquare, and the length is as large as the breadth: and he measured the city with the reed, twelve thousand furlongs. And he measured the wall thereof, an hundred and forty and four cubits, according to the measure of a Man, which is also Angel's."

When these 12.000 are converted into meters, we get a measure: 12.000 racetracks has 6.000 meters, or length, width, height =

6.000 x 6.000 = 36.000.000 x 6 = 216.000.000. It is a number that says the new city of Jerusalem, which will come down from the sky, has an area of 216.000.000 square meters. This number says that this world of ours, as well as the life of all of us, is also connected with this number (216.000.000), as well as with these numbers that we received through time, from the first second of life on this Earth, and that is: one day has 12 hours a day, and 12 hours a night and together they make 24 hours. Together in hours and different in function of hours.

One day has 24 hours:

One month has 30 days:

One month has 30 x 12 hours of day, and 30 x 12 hours of night:

I.e. 30 x 12 = 360 hours of day and 360 hours of night in one month.

One year has 12 months:

12 months has 12 x 30 days = 360 days: 360 days x 6.000 years = 2.160.000 years.

It showed us that 6.000 years has 2.160.000 days in this one information that the Lord God has shown us, we can understand that time, its length, width and height fit in the measure by which the Garden of Eden, Paradise or Jenet, or as one calls it in one's own language, will be measured.

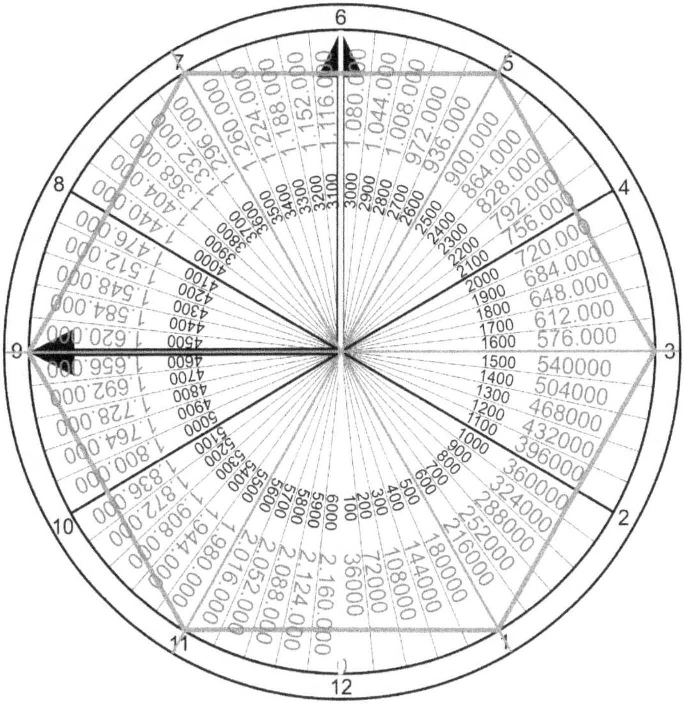

Picture 45a (SPACE, TIME, AND PEOPLE ON THE SAME GENES).

We learn that the perfect God created everything perfectly and that in the new Garden of Eden the Alpha will merge with the Omega, and that the number 360, through time and space, at that place proved its accuracy in millimeter and second, although it will take 6.000 years from the beginning to the end. This means that on this Earth, each in his own life, we have been limited by time as well as our genes, each according to its own species in which he was created or born. Nature in its kind, animal world in its own, the plant world in its own, the human world in its own. These species could mix but only with their own species, i.e.

human species by mixing it could change skin colour, character, blood, disease, health, i.e. it could exchange one particular gene type by mixing. But the genes that are the foundation of the human species could not be changed.

It is on the same principle in the animal world; each species of animal has its basic genes that cannot be changed, but in their species they can change same as humans. In the plant world, too, they can be mixed or grafted in the same species. In the natural world, such as water or air, too, genes they exist that (cannot) be changed.

And why can't they change?

Because nature, and it includes: the Sun, the Moon and all the host of Heaven, Water and Earth, have their genes on which life depends on Earth and them the Lord God created in his time. Until on the fourth day of creation, the Lord God did not allow people to manage these natural creations, because any change in them makes a great disturbance in the balance of life on Earth. Although today people and scientists are doing research and can change God's creations, which they are not allowed to change. They do it for their personal interest.

Although people think they can use God's creations for their own interest, when you disturb that direction of God which is perfect established by the creator of Heaven and Earth, then all their goals which they did will fail. When they destroy that perfect direction of movements of Water and Air and when they disturb the natural order on earth which has been going on since the beginning of life, then it will interrupt the healthy line of life and there will be a time of disturbance in nature. Everything will start

losing its order. The plant world cannot exist without natural elements such as Water, Sun and wind.
That means, the human world cannot exist without the natural, plant and of the animal world. That is, by changing the natural balance, which works perfectly, just as everything works perfectly throughout the Universe, there will be a disappearance of life on this Earth, just as Adam / Adem, of his own free will, agreed to reject God's instruction which reads: "Do not touch the fruit from this tree, nor look at it, because the day you taste it, the day you will die."
Adam did not die the same day because God felt sorry for the man he created and found a replacement for Adam's death. The one paid who did not deserve it and who offered to save the human race from eternal death. That over time will bring him back to eternal life, through the great sacrifice ,through the lamb of God who came down from Heaven and died for a sin deserved by our forefather.
But it so happened that Adam was also expelled from garden of Eden and that he no longer had access to the tree of life which he used as food. Due to the lack of that heavenly food, Adam / Adem began to die that day, because his cells no longer had food that sustained him in eternal life. Although he lived 930 years he still died, just as God had told him.
So is also this planet of ours, when its relationship with God, which God himself created to function perfectly, is disrupted, through ending that relationship that people will end, just as Adam / Adem did, of his own free will and because of his personal interest did

so, Adam had a love for a woman, and in this case it is a love of money or of domination for the benefit of one's own interests.
As a result of this event, which has already begun to happen on this planet and which can no longer be stopped, nature began to disrupt and to unfold its process, it began the same as Adam / Adem, it gradually started to die out.
Although even in that extinction it will still exist for a while, in very difficult living conditions.
Gradual extinction has occurred and will continue until the time when the Creator, God, who created everything, will no longer be able to watch the suffering, first human and then suffering of the whole system he created on this Earth.
Out of his love for all creation, and most of all out of love for Man, he will shorten the time it took to live, and he will stop the flow of life on this Earth and in this Galaxy.
Also, in the word of God it is written: "And the Lord God will come to this Earth, to reward those who destroyed this Earth." Through these words of God we come to the information that the Earth will be destroyed, in the form in which life took place on it .
This means that the beginning and the end will come together, and that it will be again beginning but in a different form, i.e. in the form that was created, at a time marked as zero and with the words, "In the beginning God created the Heaven and the Earth, and the Earth was without form and desolate, and darkness was over the abyss, and the spirit of God was rising over the water."
And God said, "Let there be light, and there was light."

We come to the conclusion that there will be darkness again and that the Earth will be faceless and desolate again. That the spirit of God will rise above the abyss, as if God would say again, "Let there be light, and there was light."

But that light will also be God's light and in that light of God will bring life back to this planet Earth, eternal life lost through sin and disobedience according to the one who created it and because of his perfection, as the perfection of all created could not fit Adam's sin and disobedience to God and to the law of God, for sin is committed of free will and consciously. Adam was not deceived as that Eve was deceived. He had a choice but his choice was disobedience to God, in exchange for fulfilling his desire and staying with Eve / Hava.

Since the Lord God is perfect and his laws are also perfect, disobedience to those laws is the opposite of perfection, and anything to the contrary creates a disturbance in the functioning of perfection. Just throw yourself out of that balance of perfection. So every transgression of the law of God is at the same time a transgression of the law of eternal life. Eternal life is conditioned by respecting and keeping the law of God. We should pay respect to God, the one and only, who created all that is seen and cannot be seen, who gave life to all that moves, who sustains the entire Universe with his wisdom and his power, and he who only says (be and it will be, who is merciful, sorrowful and slow in wrath, and whose name is love and through his love and his life, given on the hill of Golgotha for man, to bring him back to the place where he also created Paradise / Jenet or the Garden of Eden.

We need to seek God, keep his law, the Ten Commandments of God, through faith with the lambs sacrifice, sacrifice from Golgotha, who died for a sinner, i.e. for every individual who believes in him, not to perish, but to have eternal life, for all people, from all nations, tribes and languages, because with God there is no faith, all people are equal before God.

Black or white, emperor or servant, all are the work of his hands and the spirit of God is in them. All correspond to the same God and the same law, and all return to the dust from which they were created. Everyone has the same chance to get eternal life, God will not harm anyone if people do not harm themselves by their own free will. Thus, when they stand before the Creator, on the day set for it, they will have no excuse to be deceived and because they sold their eternal life so cheap.

Blessed are those whose names are inscribed in the book of life from the beginning of all of time, and who, cleansed their sins, stand before the Creator, the Lord their God, who were justified and who sought forgiveness for their sins while they were on this Earth, who pondered God, his law and those who tried to keep God's law believing that the blood of Christ and his life have the power to lighten them and prepare them for eternal life, for that is the path inscribed in the universe.

Everything visible around us instructs us that we are also part of that Universe and that it has a precisely inscribed direction that leads us from the Alpha, where we were created, and through Omega, through time and space, again to the Alpha, i.e. to begin with. But the word of God says, "The wise will understand."

Whoever seeks, he will find. Nothing comes if we don't ask for it, nor can anyone else ask for us. Everyone has their time and their free will through which every goal comes.
All those who reach the goal, all those who manage to master their own (ME) and whoever overcomes himself, will reach the sea of glass and enter the door of Garden of Eden. Their names will be written in the book of life, in the book of the lamb. And the lamb, the sacrifice of Golgotha, who overcame our death in his death, will open the door to eternal life for all those who find themselves worthy to enter that door.
Eternal life conditioned; is life in this life, so information from this life is transmitted and fits into the future life. And that is: keeping the law of God, which contains the 10 commandments of God, believing that Jesus / Isaa is our creator, redeemer and our saviour. That he is the lamb slain from the foundation of the world for Adam / Adem and all his children, for all those who through his blood and his sacrifice seeks from God their second chance to return to eternal life.
Everyone has the right to ask and that should be their main goal, which is why they live this transient and short life. But now, at the beginning of time which is counted after the first second of time creation, i.e. holy time, and on which the 6th sixth day of the week, Friday, is also the 144th hour since the creation of time on this Earth and man on it. That same hour takes us through these 6.000 years and passes through time, until the time when the future time will have the foundations that were laid in the Garden of Eden.

Those foundations are connected with this world and with the people of this world. They were before our eyes, to each individual, throughout our lives. No person, who was born on this Earth was not without this information. But people did not have time to see their time, nor the time that God constantly showed, from the first second of time. Or people did not have time to see it. Everything around us, what we see with our own eyes, shows everything the transience of life as well as the transience of ourselves.

Through time, (through the first week) of the creation of time, we see that it was composed of a life that will be lived after this time, i.e. with the Garden of Eden sheltered from sin and which again will come to the same place where it was in its first origin. It will have the same dimensions as well as the shape it had in the first creation, and is associated with the number 144. The wall around the Garden of Eden or the new city Jerusalem, has a width of 144 cubits. And the wall is tailor made to mankind which is also Angel's too, for the people who will enter into it with the lamb will be Angels.

And these are the people who will come from this Earth and who will have the form of Angels, for this body, which God created from the Earth, will remain on the Earth, and the spiritual Man is the one whom God will raise from the dead, i.e. Jesus, who has the power given by God to be able to create us again and revive us when he comes on the clouds, at the last moment of life on this planet. At that moment of the merging of Alpha and

Omega, when eternal life returns to its eternal line, because it will be returned by the giver of life himself, and that is the one who has the power to raise the dead from the graves, for only he has the power given to him by heavenly Father, God; his firstborn son in Heaven, and that's why he came to this Earth and was born in a human body and who died in a human body, returned to his Father and our Father, and through whose death we can justify our sin and get eternal life again, it is he who raises people from the grave and leads them into eternal life and the new city of Jerusalem.

The 144th hour, at which the Lord God created man, would be the measure of the wall of the new city of Jerusalem or the Garden of Eden, which is the measure of the Angels, which is also the measure of Man. He showed us that through the time that we live on this Earth, that it will also be the measure by which we will measure or weigh ourselves as we enter the Garden of Eden. The name itself (Jerusalem) says that it is (Alem or Jasper) and that on it is the foundation of eternal life. And the building of its walls was Jasper or Alem, and the city was pure gold like clear glass. "And Jesus says that He is the cornerstone, the Jasper stone." On it everything is created, stands and exists, and that cornerstone holds everything visible and invisible by his power and wisdom, and that it is through him everything that came into being and that through him everything that came into being survives, because He is the cornerstone on which everything rests.

Like the first page, the first book, written by people who God endowed with his holy spirit and to whom He gave instructions which they wrote, to people for science as well as for the instructions they need to live. They contain everything you need to know, and the answer to every question related to life on this Earth as well as to the life that will be after this life, as well as to two directions leading to two completely different lives. One direction, leading according to the Creator of all creation and whose throne has always been on heaven and who will descend to this earth at the moment when the re-established, sinless life that is the Creator returns who gave it to Adam when he created him. That holy life is being renewed as well as everything that has been renewed during these 6.000 years that have shown the direction of movement of all creation.

It is written in the holy word that the throne of God will descend to this Earth as a memorial of the great love of God and the great suffering of God which he had towards the fallen human race. That he also gave his firstborn son through a plan of salvation, so no one who believes in him perishes but has eternal life. Through His son He also gave Himself, through His love and through His suffering to the sinner. And another direction that leads in the opposite direction from the Creator of everything visible and invisible. It is a path that remains only on this planet Earth, which has no exit or door. And all who follow him are prisoners on this Earth. Because they didn't read the road sign before going on that one path, they brought themselves, through their free will, to that path without exit.

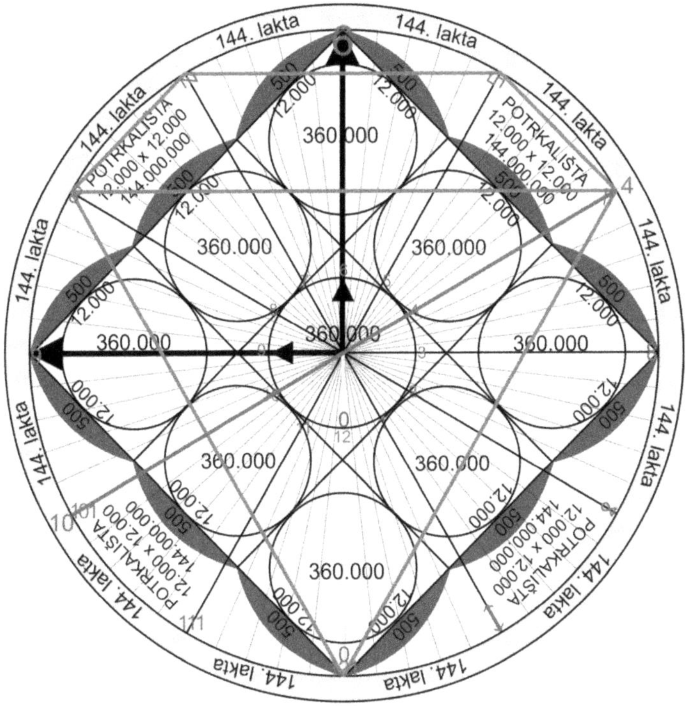

Picture 46 (FORM OF THE NEW CITY OF JERUSALEM).

It shows the shape of the new city of Jerusalem, as it stands in its Alpha and Omega, in its space and its greatness. Just the way it is shown by the Angel to the Apostle John, a disciple of Christ / Messiah. The city stands on four corners, its length, height and width are: 12.000 x 12.000 racetracks, and length, width and height are time, space and a number that fit with the length of time this planet has, the place where it happened at that time and the exact number of people who will also enter the Garden of Eden. People from this earth who, at the moment of entering eternal life, will have the form of an angel, because they will be

justified of all their sins because they sought that justification from God and kept the law of God and had the faith that the holy blood was spilled on Golgotha for them.
And just by their deeds alone they cannot earn eternal life, but the justification of their sins sought through their Creator, savior and through their redeemer, Jesus / Isaa. Only he was the lamb slaughtered, from the beginning of the world.
The wall has 144 cubits of man, man was created in 144 hours from the origin of life on this earth which also is the (measure) of Angels. The city has 12 gates, 12 grains of pearl. Each gate was a on grain of pearl, and the streets of the city were pure gold, like glass too bright. And the 12 gates are 12 hours in one day or 12 months in one year, by which time is counted.
There will be no Sun or Moon because God's light will illuminate. These explanations complete the description of everything we need to know and what is related to the instruction and direction that leads to eternal life, to our Father, to our God and to our Creator. Who showed us all his signs, so that no one dies but everyone who, of their own free will, wants and who wants to be on God's road, to find God's way and to stay on it, with the help of God. That is the only goal of our existence. Everything else has no value because of which we would miss our chance we got through our time on this earth as passer-byers.
Below is a brief description for the purpose of better understanding about man as a part of the Universe, as a better belief that we are truly created in the image of God as God told us he created

us. While living in the Garden of Eden, Adam and Eve had the image of God, in addition to the image of God, the holiness of God, because they lived in Eden as holy beings. The moment they sinned, they both had to leave that sacred space. They could not use a space that belonged only to holy beings.

The holy word of God proved to us in the Garden of Eden there were all kind of animals that enjoyed it with them in that sinless environment. And among them was the snake, as one of the most beautiful animals. Being the most beautiful animal and attracting attention with its beauty, Satan - the Devil took advantage of its beauty and used it.

Even today, many beautiful things attract our attention and distract us from God, leading us in the opposite direction from eternal life. From this information we see that Satan - the Devil used the serpent and her beauty to reach his goal. And the goal was to deceive the man and to separate him from the one who created him and then use it for his own purposes against God.

Eve, did not listen to God's warning not to go to that (tree) or look at it, because by doing so he will create a temptation that she will not be able to resist. Despite the warning, she went under that tree. And it (happened) what (happened) and lost her holiness and the right to further residence to the sinless world, for no one can work against the legislator, the Creator, and be under his protection again. He immediately loses the right to protection. This happened to the first woman, as well as to the first man, who out of curiosity entered a world in which they should not have been. But the consequences of their disobedience spread to all their children. The one who created them gave them like us, a second chance a return back to a sinless life.

When the Lord God determined the reasons for their disobedience, even though he knew exactly what had happened, he wanted to hear them switch their guilt on the other and them justify themselves. God has given people free will to do what they want, but through their free will and their answers for their sins they will be responsible when a certain time comes. That's right, Eve/ Hava said that she was deceived by a serpent, and Adam/Adem said that he was deceived by a woman whom God associated with him.

Everyone shifted their guilt to others because, after they tasted sin, they immediately began to lie. God knew of the consequences they had caused by their disobedience and that they would have to die, and that they would lose the form of life in which they had lived, he would pity them and make a plan of salvation, to bring them back through him, to the place where he created them.

And through him sacrifice and through his love he postponed the time of death and gave them a second chance, in justification through faith. But regardless of his love, everyone, individually, heard immediately for the consequences that resulted from their disobedience, such as the fact that they would have to die. But even because of them, someone who has the strength should die to bring them back to life. After leaving the holy life, at the gates of the Garden of Eden, they slaughtered an innocent lamb and immediately offered a sacrifice for their sin. And it hit them hard.

The penalty for sin is death, and someone had to die for their sin, although they, too, died. But they could not, with their deaths justify their sin. Only the one who had the power could do that, to give life, i.e. the one who created them. Before they left the Garden of Eden, the Lord God told them the punishment they deserved by breaking his law. Adam / Adem got his punishment, Eve got her punishment, but the snake, too, got his punishment. Those punishments were also their curse. The snake received a special curse. Lord God he said to him: "Because you did this, More cursed shall you be than all cattle and all the wild beasts: On your belly shall you crawl and dirt shall you eat all the days of your life." At that moment, the snake lost both its appearance and its beauty, all what was left, was its spine and head, and it is still dragging to this day. On the snake, its skin and its shape, which it got in the hour when God cursed it, we see it dragging itself on the dust of the earth. But apart from that shape, cubes in different colors and in different shapes can be seen on its skin. And in those places where it touches the ground, there are different patterns, which are not in the shape of cubes or rhombus cube. Its color changes in relation to the environment in which it lives and according to the season. In its curse, God shows us one very important information, and that is the snake in translation means: z, mi, ja; which means: we, I, the snake. We are all snakes! And since we can all be snakes, we will get an answer from the Universe.

Since the snake has lost its form due to a well-deserved curse and since it continued to drag itself in the dust which shows us that God cursed it because its beauty was used for a curse instead of blessing. This contributed to the deception, through its body was used against God and against God's commandments. It received a special curse that it will have until the end of the century, as proof of the truth of God's words.

Through it, the enemy of God and man, Satan - the Devil, tried to create his kingdom on this Earth, independent of the one who created him, independent of God and stood opposite to God. Thus the Serpent became (the medium) through which Adam and Eve were deceived.

And they were thrown out from the part of eternal life, from the Garden of Eden, and through the serpent, as well as its body, all the torments for human beings arose as well as for the complete life of all creation. Because all created lost its sacred form and all began to die, as did man. Cubes can be seen on the snake, on its skin. God has on those cubes a message for all people:

"Take a look at you yourselves!"

We, too, have the same cubes on our skin as the snake, only our cubes are not so conspicuous because they exist in similar shades of color on our skin and it is difficult to notice due to the shape of our body. We wear clothes and our body is covered, and the patterns are not noticeable. Just like the snake distracted Eva's attention.

But when we look in the other direction, the opposite of our physical beauty or physical appearance, then we will see that cubes exist on our whole body, just like a snake. Where the cubes are shown in these places, at the same time, there are elevations on our skin or recess that separate the cube from the cube, depending on its size located on that part of the body.

Our whole body, the outer part, as well as the inner, i.e. all our organs are composed of shapes that show us that we are in the shape created by God. And that means, that we dependent on our creator, our God and that we cannot exist without him and his power that keeps us in our existence.

Just as he holds the whole Universe.

It also shows us that we are made up of the same particles, elements and of the same matter, as people, and nature and all that there is; all in its form and in its elements, of which is composed, shows that everything is of the same composition as man, only in another purpose of existence.

When we start from man and his lines of life, which are in their own existence merge with the whole Universe, and which in that merging into lines are connected with God, who created them, and that they do not serve for beauty or any insignificant purpose, the Lord created them so that through these directions, which are various geometric shapes, through them we see the whole universe in to one unity and in one harmony that exists and lives beyond itself of one source of life, and that through God who created it all, so perfect.

Whose strength and power is at the center of everything and through God's heart and of his beats, the beats are transmitted to the whole universe, and to the beats of all created beings and to Man. And at all times has direct connection with each creation.

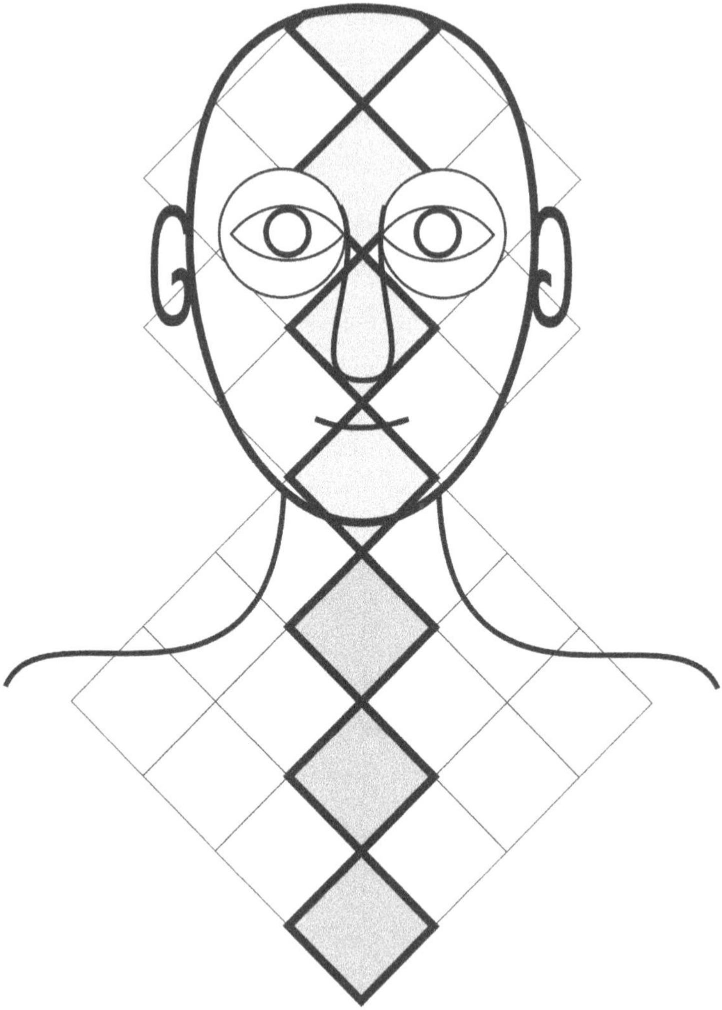

Picture 46a (MAN IS PART OF THE UNIVERSE)

It shows man and his lines through which he is connected with, the Creator God, as well as with the Universe.

Starting from the head, cube-shaped lines are drawn directly. And always a cube that is slanted and starts at number 6 on our

clock of time, which we use in everyday life, and which is also an indicator that time passes as we do with time.

Only by the hour, which shows that one day has 12 hours and that night has 12 hours, that is the middle of the day and the middle of the night are at the 6th hour. And that 6th hour is very important. Because in the moments when Jesus died on the cross, on Golgotha, there was darkness from 6 to 9 hour. So it says in word of God. When the line is drawn from 6 to 9 hour then it goes diagonally. The new city of Jerusalem or the new Garden of Eden, too, has a slant appearance. It is also a cube that has four sides and three doors on each side. Then, all these cubes that have four sides in appearance, change their appearance, depending on which angle you look at them from, in fact, both the cubes and the rhombus cube are at the same time.

Starting from the head, over the forehead, face, neck and all over our body are the same cubes and patterns as in snakes. Our skin is covered with the same cubes, in smaller or larger sizes, but the cubes change their appearance into rhombus cubes, also on a snake, they are sometimes cube-shaped or tilted, rhombus -shaped. Also all our organs and their forms are in the same shape that the Snake has. Its forms, both in appearance and in the structure of the body, correspond to the structure and appearance of Man and all living beings. And they are circles, cubes, triangles or hexagons. Our body, inside and out is covered and composed of geometric shapes. These forms 100% fit into each other, through our body, as well as through the body of the snake, as well as through the body of all living beings.

And that means the snake, which lives without its form, which it lost because of his curse, serves Man as living proof that we are all one whole, from people to animals to nature. I.e. every particle that there is has a geometric shape on this Earth that has everything created, both on Earth and in the entire Universe.

The Lord God created Man from the Earth, and the Earth is also part of the Universe and it has a form that has the whole Universe. And these are 360 ° circles and on those circles, which are called stars and the planets and which were created for life on this Earth, are an integral part of our galaxy.

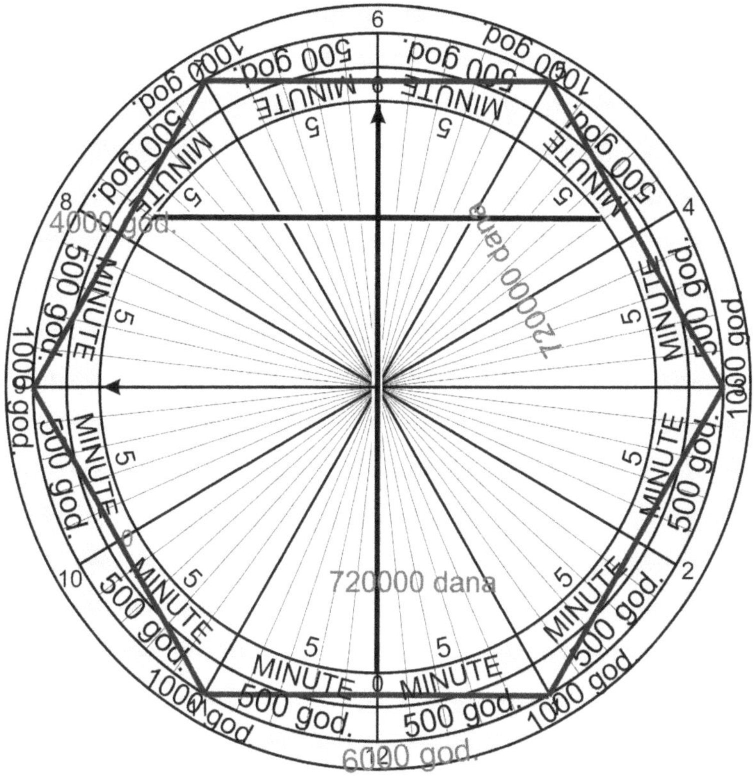

Picture 47 (LIGHT, LIFE AND SPACE THROUGH TIME)

It shows the shape of time and space or the shape of Alpha and Omega in space and time of the whole life of all creation, both on this Earth and in the whole Universe. It confirms the accuracy of the information, that God created the nature of the same shape and the same matter as Man. These pictures confirm the evidence that shows the seals and they have sketches of hexagons, as Saturn has a large hexagon at its top. All the bodies of heaven testify that they are connected by the same force and that they were created by the same creator - the Lord God.

All heavenly bodies testify that they are connected by the same force and that they were created by the same creator - the Lord God.

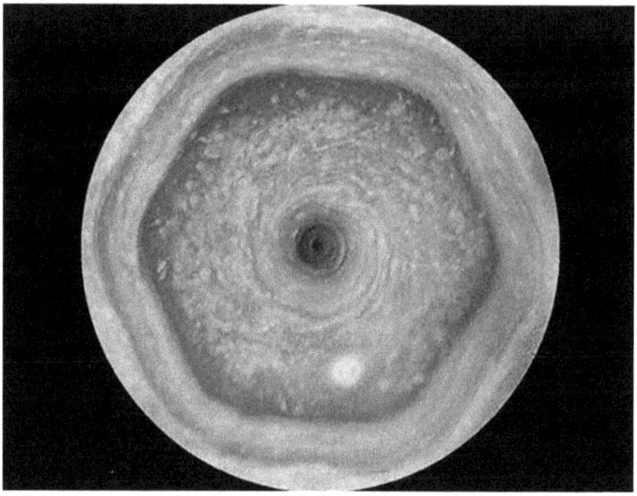

Picture 48 (SATURN AND ALL MILITARY HEAVEN-PART OF LIFE UNIVERSE).

In winter, snowflakes that have a perfect hexagonal shape sometimes fall, and on that perfect shape they show that when we connect these arms with a line, arm by arm, we will get a hexagon. Then, the bees in their God-ordained wisdom, and us as a lesson and revelation, they show in their work, in formation their spaces for their needs, in their sachet and wax, make a perfect hexagon.

All this information directs us and tells us that nature and the universe serve as a guide and direction in which we should go. We can affirm to ourselves, as well as to others, that nature and time testify to being on number 6 played the most important event in the entire universe and that the creator of the universe Jesus / Isaa, just at the 6th hour began to exhale and that he had his head at 6th hour, which is also the middle of the day or zenith or noon.

But even though it was the middle of the day, the Sun lost its shine and there was darkness from 6 the sixth to the 9th ninth hour, for that is who created the Sun sheltered its radiance with his head, in his mortal hands. These events, which are confirmed through space and time and which is confirmed by nature, as well as everything created on this Earth, serve only to make people understand the truth and come to the truth through the truth. Find the path that leads to eternal life. Number 6 is at the same time a Man's number, for Man was created by God in 6 the sixth day, on Friday. And after 4.000 years on the 6th day, on Friday, he died for man that he created , who confirmed it all and gave it all information that serves to confirm themselves, that anyone who is looking for truth and life not to wander, but to find it and live forever. From these information described above, we have come to the information that Man on the inside, as on the outside, is composed of sixes, and that everything created has the number six on it.

In the last pages of the Bible or in a book called "Indir" or The New Testament it is written in the Revelation:

CHAPTER 13:

Verse 18. "Here is wisdom. Let him that hath understanding count the number of the beast: for it is the number of a Man; and

his number is (Six hundred and sixty six) or 666."
This information says that man's number is 6, and that the number of the beast is the number 6. That Satan's number is 6 or 666.
But number 6 indicates that it is Man's number, because man was created on number 6. But other information says that Satan - the Devil deceived Man for the reason that he came to his throne through him, for the reason that
he wanted to have his throne, independent of God, and what he wished to be God, that all creation belongs to him.
In him envy and the desire to be the king of the universe, he thought he could make it happen because he got great power, which he received from the Lord God who created him.
Since he was obsessed with his power, he forgot that he has that power from God; but he tried, through Man, to fight for his throne.
At that moment, when he came to Man to sin, and when man, because of disobedience to God who created him, had to leave his sinless homeland, he lost everything that was connected with that form of life. Although his death was postponed for some time because God found a substitute to atone for his sin, he still died.
Satan, too, began to die because he too was cast out of sinless life and his time of dying is still going on, because everything has its own time.
And he is tied to number 6 because the days of his life are limited and they are at number 6 or 6.000 years. And 6.000 years is also the time God gave to this Earth and the people on it. This means that people, ever since they have lived on this Earth in this sinful

form of life, since time counts, they have lived for almost 6.000 years. Satan, since he deceived Man and his time is counted from that same moment and he is at 6.000 years. So is Satan and Man's number 666. He has three sixes and from the first moment since he deceived Man, continued to deceive him and will deceive him as long as there is life on this Earth. But from that moment on, he continued to cheat people in all possible ways. Every deception of Man goes to Him for his benefit because even now he counts on the number of deceived people, for whom God does not want them to die an eternal death, to (benefit) himself. He thinks that is his salvation because if God justifies a sinful Man that He too should also be justified.

Because of sinful people, the one who created them came down and died for them. But because of the originator of sin - Satan - the Devil, He did not come for Satan; he has his time of grace of 3.000 years, exactly as long as Adam lived in the Garden of Eden, in the sinless life. He did not, for that time of his mercy, seek the mercy and forgiveness of his own sin. But it is on this Earth, by the deception of Adam/ Adem who was also the king of this Earth; fully took his crown and became King of this sinful world, and from that moment he deceives all people. First the time was changed that people may not know exactly what time they live in. Then he changed the order of the days of the week, so that people do not count the days exactly as God created them. And the seventh day of the week, which is the Sabbath, it lost the significance of the holiness that God has given it.

He changed the hours on the clock and turned them the other way around
the direction of time, as God has ordained. And that is: And it would be evening, and it would be morning - the first day. Time goes opposite of God and shows the hours on clocks that move in the opposite direction from the time that there is. And in those hours they show the direction in which the middle of the day is at 12 o'clock, instead of 6. It changes everything so that people can't in their speedy way of life in which they live, seek and explore in which time they live in. Nor are they interested because they are preoccupied with transient things that seem important to them. Then, at number 6, he made his throne and tried to claim it as his number.

He can do all this, up to the time he has before his (disappearance). He is a lie and the father of lies, because he told the first lie, and everything that is created on a lie cannot last long. Everything that the Devil has - Satan, he stole everything from God, and set it up as if it is his creation. All the numbers he uses wants to rewrite as his, it's all God's, because it's all God's creation, just like he is. And all of God's creations have their time of existence. Depending on what God created them for. But if a creation is removed from its function, it cannot perform the function for which it was created. So by deceiving the Man, he deceived himself as well.

For man; The one who created man gave his life, and through his life, gave life, to everyone who, by his free will, wants eternal life. Our veins are long channels and in their direction, in which they are formed, they look like snakes. But they are therefore shaped like a circle through which blood passes. The internal appearance of the veins is covered with hexagons, all next to each other, i.e. chained. Hexagons have the same look as well as wax that bees make on their hive or that is on Saturn, and which is found in every creation of God.

The same shape of the hexagon, which should show us that it is also the main (gene), on which is everything that God created. Same as through all this time, from the moments of the origin of the first minute, the first hour, the first middle of the day, from the end of the creation of all that God created, and all that he created, created in 6 six working days.

As it measured, 6.000 years of sinful life on this Earth. All this information was shown to us by God, so that we would not have dilemmas about the truth he affirmed in all creation, what we see around us, to guide us on the path of truth and eternal life. On that number 6 or on the 6th hour of the day, or on the 6th day of creation on Friday, or on the 6th day of the week, the creator of life began to die, for life's of the sinful people, for everyone who finds the way, with God's help, and returns again to eternal life. The hexagon also shows us the number 6 in which everything is created and on which on Friday fulfilled the plan of salvation given by God and that from 6 to 9 hour our savior died, so that we could live. It is in this, our veins, which are composed of hexagons, our blood passes. And blood is the life of every man.

Our blood has the form of Alpha and Omega, i.e. 360 ° circles. These circles, which are an integral part of our blood, on which our lives depend; are
(erythrocytes), and they are the main carriers of life in our body, and their shape is a (circle) or Alpha and Omega.
Jesus says, "The life of every living being is in his blood, or blood is life." From this information which show that everything we see and don't see has God's genes which connect to each other and which are emerged from the creator, and exist because he feeds them with his own energy and his life.
This (hexagon) has everything created in its form, just like Man, and it is, at the same time, the (gene) that binds all created with God. Because on each leg of that hexagon there is also extended line connecting it to the next hexagon. And that direction it goes to infinity. We are not our own and we belong to the one who created us, who shows us the way to all his creations, the truth and life. Through all this information, which confirmed that it is the lord God that created everything visible and invisible, as well as life on this Earth, who also created us and who revealed to us that everything was created equal in time and space, and that it is equal in composition with the elements from which it was created. It is the same in the appearance of the form to which all God's creations exist, regardless of whether they belong to living creatures or creatures that serve the life of living creatures. Every creation has the same appearance in their creations and the same composition of matter from which God created, same as man.

Regardless of the fact that some are seen and some are not, they are again in their creation on a common basis, they are established and connected in their function. From these common functions they depend on each other. They have one and the same rhythm that God gives them and that keeps them in existence.

All creatures are made of minerals that survive thanks to light, which has power, through energy, which is created in light by the power of God, through which the life of the entire universe is moved and maintained, with elements that are perfectly aligned in their functioning through the energy of electricity, magnetic forces and chemical compounds. Using all these, united God's perfect creations, they produce the life of the Universe, as well as the life of people and the whole life of this planet of ours, which is also an integral part of the great Alpha and Omega, in which the life of all these creations is created and renewed as well as the growth and expansion of the Universe.

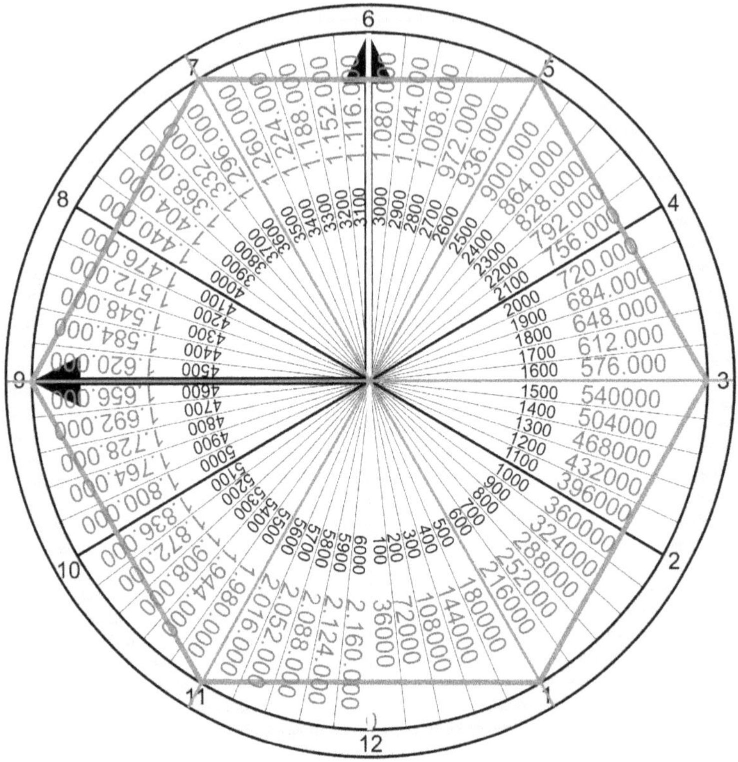

Picture 48a (TIME AND LIFE THROUGH ALPHA AND OMEGA)

It shows the hour that takes us through the time of all these 6.000 years and lead to the end of time. It also showed the time that Adam and Eve lived in a sinless life, in the Garden of Eden. It also showed the time at which the birth of Jesus Christ / Messiah was fulfilled and on which, at the same time, the counting of time (old era) ends. On which the plan of salvation is fulfilled, and the time in which Adam's sin would be paid and in which death, in the sacrifice of Golgotha, turned into eternal life for all who seek the path of eternal life.

It showed all these sixes, which the Creators shows us, and beyond which proves to us that something very important is going to happen, and directs our attention to what message is written in that number. The first number 6 announced that God created Man on the 6th day of creation and that on that day, which is Friday, he finished creating life on this Earth, because he had already created everything before Man. The second number 6 announced that it was on the 6th hour (noon and midnight), because there are 12 hours every day and the middle of the day is at 6th hour. It is also mid-night.

The third number 6 announced that God, on Wednesday, which is in the middle of the week, and at 6 the sixth hour, created the Sun, the Moon and all the host of Heaven. From that moment the Sun replaced the light of God that was in those first three days and from that moment, all creation uses the Sunlight that will exist until a certain time in which it has in its existence.

Fourth number 6 announced that on the 6th day of the week Friday, Jesus Christ / Isaa / Messiah died, for everyone who believes in him, that through him has an eternal life. Fifth number 6 announced that Jesus Christ began to die at 6th o'clock at the zenith of the day.

In the word of God he writes: "And there would be darkness from the sixth to the ninth, (6 – 9) hour. Because the Sun also shielded its light from that horrible sight."

That is. The Sunlight could not be seen in its radiance because Christ's head was standing exactly on the San, at its zenith, in the middle of the day. At the place where the Sun took over God's light, on Wednesday, that is. In the middle of the week, in the middle of the day, at number 6. The sun could not even shine because the one who created it, his head, in mortal torment; fulfilled the plan of salvation for Adam / Adem and his sinner children, at 6th o'clock and at exactly the same time, after 4.000 years the one who created the Sun was born and died. And in his death showed that he is the life and death of all forms of life, both human and everything that is seen and not seen and the whole Universe. Sixth number 6 announced that at 6.000 years, this form of life in which we now live will cease.

This information speaks to all of us: who created us, who redeemed us from eternal death and who will revive us and awaken us from a mortal dream we must go through.
With the difference that one will return to a sinless life, to the Garden of Eden, to that place to which God created Man; and others will end their existence on this Earth and return to the eternal dust from which they were created because, they did not use their time properly.

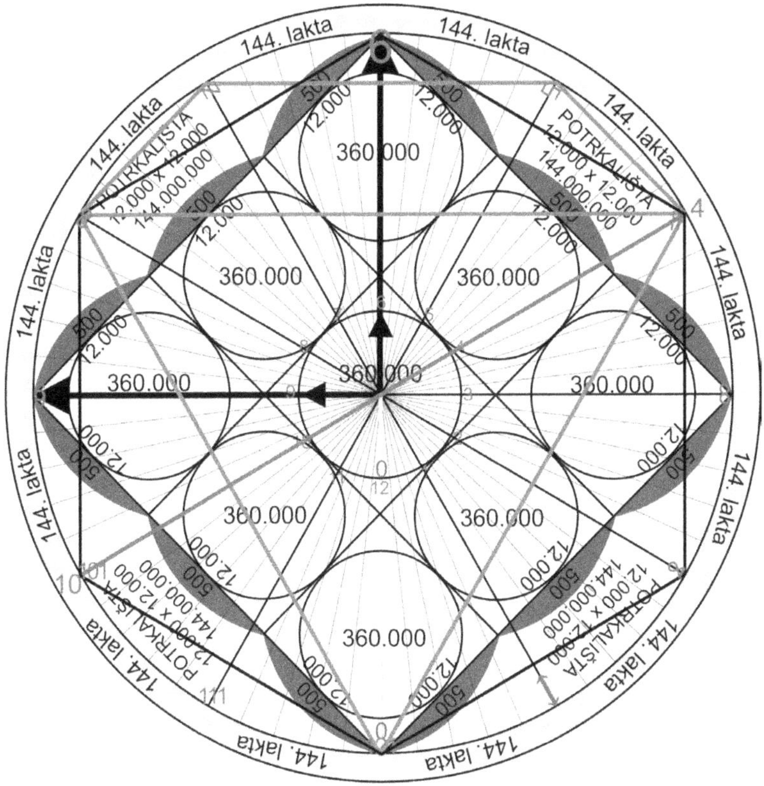

Picture 49 (12 DOORS THROUGH 12 HOURS).

It shows all 9.000 years. Time from the moment of its creation of life on earth, from scratch when the Lord God began to create life on it. He then shows all 12 of the most beautiful gems found on this Earth, on which the walls of the new city of Jerusalem were built. It shows the 12 gates of the new city of Jerusalem through which the saved from all over the world will enter from the East gate of the three , from the West gate of the three, from the south gate of the three and from the north gate of the three.

He then shows space as the surface of the Earth, which the sons of Jakov/ Jacob inherited according to God's promise given by God to Abraham and Izaac, Jakov / Jacob, and who entered after slavery in Egypt, where they were 430 years. That same space the children of Israel are counted as the measure that the new city of Jerusalem will have. The city will have 12 gates with the names of 12 sons of Israel and 12 names of Christ's / Messiah's disciples the apostles.

On it and in it, you can see an area of 9 times 360.000 thousand. And 360.000 are days that are in one thousand years. The new city of Jerusalem, stands on 4 corners and has 12 gates. When the door is divided, each in its place, because they have 3 doors on each side, we get a picture and a cube of 9.000 years. Here we see time calculated in two dimensions. It's time calculated from the origin of life on earth, it is all together 3.000 years of Adam's life in the Garden of Eden and 6.000 years of life from the origin of sin and from the counting of life in which one lives from the moment Adam began life as a sinner. In the same circle of 360 ° there is a hexagon showing 6.000 years and a cube with 9 sections showing 9.000 years. In the middle of the circle is located time that shows the accuracy of the time that has elapsed and that was counted from the first second of life on this earth.

Also, in this picture we see a jasper stone on which stands everything that God created. Jasper is the cornerstone on which everything rests.

In the middle of the Jasper stone stands a cross and on it Christ / Messiah, his head at 6 o'clock, in the middle of the day, and his feet at zero, i.e. at the beginning. The universe and everything created showed us who created us, who our redeemed is and who will create us again. In this picture we also see the fullness of all creation. Through it, God showed us everything we need to know and what is important for our salvation. It shows everything from the first particle (Atom) that functions perfectly in its creation, as well as in the conversion of energy from one whole of perfection into the infinity of transmitting the same energy which has a return direction and which is not lost at its center. It shows the connection between space and time as one perfect whole that functions in a certain time for a certain time, the time for which it exists. Time and space do not exist if they do not have their function for which they exist. There was no time or space on this Earth until there was life on it.

In the Word of God, it says that the new city of Jerusalem will have a size:

scaled to 12.000 x 12.000 racetracks. Height, width and length the same. And to stand on 4 corners.

Or 1 year has 360 days:

10 years has = 3600 days:

100 years has = 36.000 days:

100 years has = 36.000 : 3 = 12.000 days.

1000 years has 360.000 x 9 = 3.240.000 days.

Time of 9.000 years + 1.000 years of consecration of the Earth = 3.600.000 days. It is also the time of the perfect whole of the universe on which time, space, life and truth was created. Where there is no more lies.

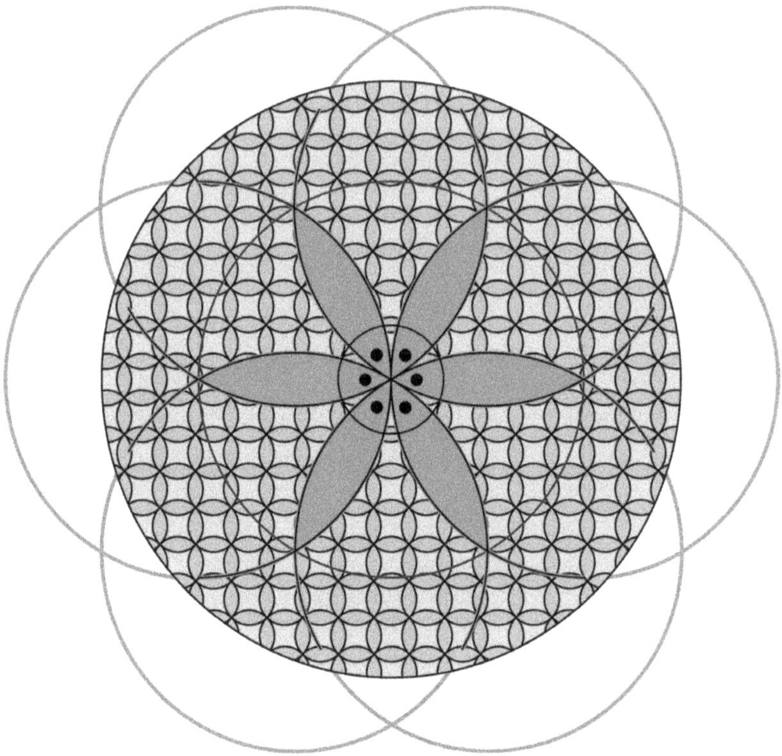

Picture 50 (GOD'S POWER AND LIFE ON THE SAME LINE).

Geometric shapes can be seen in this picture, which is in fact a confirmation

that in one whole, which is the creation of God and which was created to carry out its task assigned to it by God.

That is, it is one of the thousands of Alphas and Omegas in which is fulfilled and completed the perfection of this great Alpha and Omega that surrounded our galaxy called the Milky Galaxy. Where they will soon connect the end point, (O) with the starting point (A) and end every action and every life that took place on it for a certain task or function assigned to it by God.

In this great Alpha and Omega, it can be seen that it is composed of thousands special creations that were each separate in their function of Alpha and Omega.

Also, it is seen that in the origin of life and the origin of light, first God's then the Suns, chemistry or chemical processes were created, which with their energy maintain the vital functions of everything created. These are, in fact, atoms or atomic particles that are also circles of 360 ° and which by their unification create great energy. Exactly as much as is needed at that distance, from the Sun to the Earth has a 100% positive effect through the action of that energy on the creation for which it was created. I.e. to move life through his own light and energy in various forms of existence.

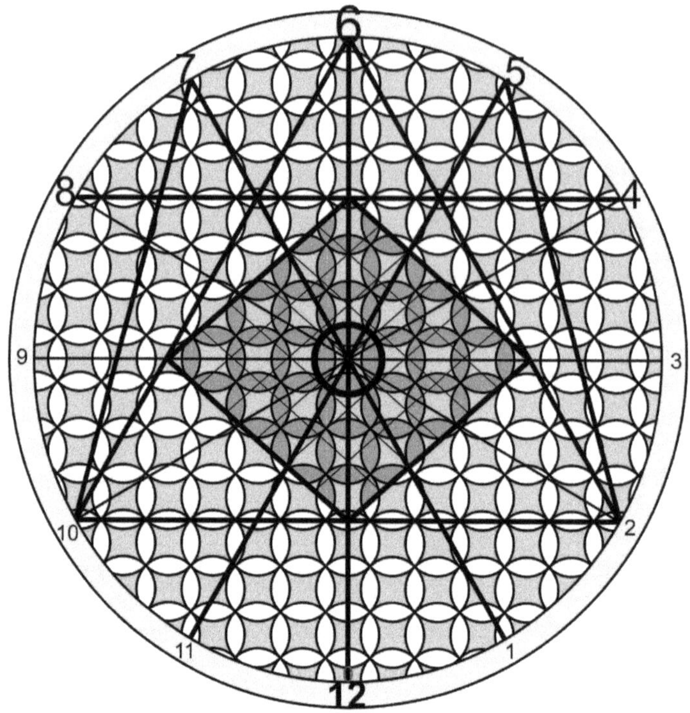

Picture 51 (THE POWER AND ENERGY OF GOD IN THE SAME PLACE AT THE SAME TIME).

Then you can see the lines that first lead from the new city of Jerusalem and where it says that it is a cube and that it stands at 4 corners. That its height has equal widths and lengths. And that means that its appearance is a cube, and that from that cube continue all other cubes and continue on top of each other, to make a Universe that looks like a cube (in) a 360 ° circle. All these cubes multiply through each other and neither the start nor the end can be seen. Because everything created, God created so perfectly that it was not noticeable when the Alpha and Omega came together nor is it felt the moment when a new is created in the contact of two points. Although they have two different functions in their existence and each of them is, also, separate Alpha and Omega. Again, like all other Alphas and Omegas make the process of life and death and life again. Although people every day, that is, every night, when they fall asleep, they lose sight in, the length of their sleep. On the same principle, the Creator created the awakening after death.

No one will have a sense of how long he slept in that death of his. There is only a difference in that awakening. One will awaken to eternal life and the other to eternal death. Depending on their free will that they used in life while living life in this time, on this earth, because eternal life is conditioned by life on this Earth and in this time, which after this time is transferred to time in another form, which is also tailored to the space on the new Earth.

We have within us the same creations from which we are made and we all exist thanks to the power and wisdom of God. And that power and wisdom of God is the foundation of the life of all creation. God has created all the conditions in which life can be maintained, both of nature, animals and man.

And God created man in his own image. All the creations of God testify that we were created by a Creator who has power and wisdom because they are all perfect. And everything exists, complementing itself in its existence, and nothing created could exist only in its creation and only for its life function. Just as our organs could not function and exist if they were not connected to each other. And if they would function, independently of the other organs that complement their function, then that function would not have the same effect because it would not benefit the whole being, but would only do work, and it would be useless work. But when it performs the process of one specific goal within other organs goals and when it is constantly working for a certain goal or task, then it is constantly working for a positive process on which all the organs of the body depend on one another, just like that body that would be separated, it could not exist by doing a function only for itself.

So we see that our body has the same meaning and the same work process as the whole universe. All the creations of God work and exist so that all creation has a common life in that creation. And nothing exists that exists only for itself, because life consists and exists in the common sharing of life and the vital needs of all the elements related to each other. God created man in his own image. He created nature and everything that exists on this Earth and in this galaxy. Everything created testifies only for itself, that everything was created through the power and wisdom of God, every animal that moves and has the breath of life in it, it is a living witness that it is part of the Universe and our galaxy.

On every animal, as well as on man, there is a seal of God. All fish in the sea have on their gills and on their protection, which is located on their shells, drawn cubes in various positions depending on the environment in which they live. A crocodile on his skin it has a seal that the galaxy has, a hexagon. The chicken on its own legs and skin has the same sign as snakes and humans. Our world has on its cracked crust, like the skin of people, the same, identical signs that are small and large cubes and rhombus shaped, connected to each other and connected throughout the whole surface of its expanse. This same chain of connection is composed through living beings as well as through nature and confirm that we are all creations of the same power of God and the same matter because we are all of the same matter or the same composition on which God established everything that exists on this Earth as well as in the Universe. And that we are all dependent on that power of God and that nothing can exist without it. And that everything became through it and that it can only exist through it. And that this power of God, just as Adam/Adem in the Garden of Eden had eternal life that God gave him in creation itself because he was created in eternal life, and that eternal life was conditioned by obedience to God and (food) which he ate like a sinless being. That food was found on a tree called the tree of life. After leaving Eden that tree ceased to be his food and immediately the beginning of the death or extinction of Adam's cells occurred, which gradually led to Adam's death. But to all people for guidance, the Creator also showed us his truth when he said that he created man in his own image, he confirmed every word he said.

Adam immediately saw a snake that continued to live nearby and who always had a chance to bite a Man. This means that the serpent, since its fall into sin, had always posed a mortal danger to any Man who did not take control of his security of life.

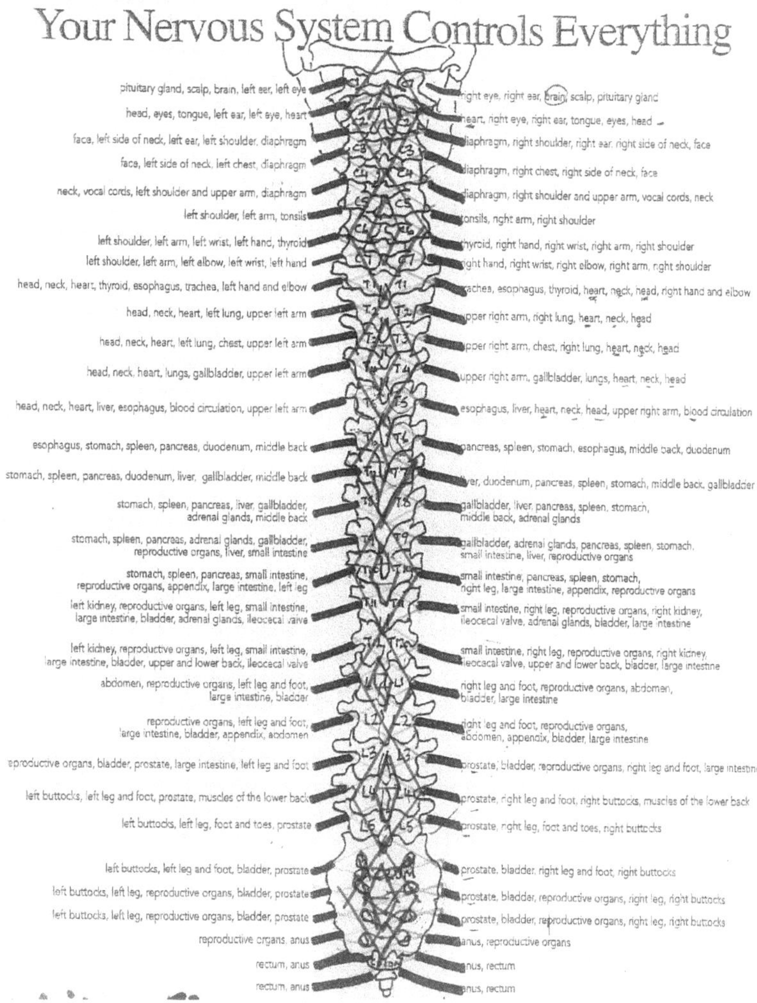

Picure 52 (MAN PART IN THE UNIVERSE)

When the sons of Jacob came out of Egyptian slavery after 430 years and traveled through the desert, they were bitten by deadly snakes. But God told Moses / Musi to make a pillar and hang a serpent on it and everyone who looked at it with faith, will survive, i.e. he will not die. All those who looked at it with faith they survived.

What was the secret here, for surviving a snakes venom?

the following information says:

The snake does not have the power to make anything good happen through it, but it does have a symbolic meaning from the day Adam / Adem saw his body naked, from the moment he lost his holiness, and the serpent from that moment lost his body he had as the most beautiful animal in the Garden of Eden. At the same time Adam, Eve and the serpent lost their body shape they had living in a sinless life and from then her naked body has shown that we all have a foundation of life or a tree of life planted in our body that tree of life, which God created and on which we exist or we live as living beings, is, in fact, our backbone, which extends through the length of our body and over which we function without which we could not even exist. Through the backbone, which looks like a snake, but it is actually a tree of life that God planted in us and that can only live depending on God's energy sources and food.

Since our spine takes food from God's sources and we take life and strength from our spine, our spine gives life to every organ in our body and is connected by the nervous system to each of our cells, and through it and through the energy and strength that sustains us, we have the abilities of life's functions.

And why is it said that our tree of life uses strength and food from God's sources?

Because all the elements, which are of vital importance, are created in the days of creation in which God created everything in his power and in his light and that was until the 4th day of creation, until Wednesday, because on Wednesday the Lord God created the Sun and the Moon and all the host of Heaven. And that means that our spine uses the nutrients that are created in the light of God until the 4th day of creation, and that was the light, water, sea, land, nature, fruits and vegetables, as well as grass, which is also what animals eat.

And that is where all the life elements that sustain the life of our tree of life or our spine are located. So our backbone is, at the same time, part of the creator who created us, his form is found in all the creations that are the main elements of the life of all living beings. After the 4th day of the creation of life of this world comes the 5th day, in which God created (life) animals in water, and on the 6th day God created (life) of Mammals and Humans.

Humans, like animals, live by using creations created by God in the first half of the week, in his strength.

And the second half of the week is Thursday, Friday and Saturday. In them are created living beings who use God's creations and his power. All living beings have used, since their inception, Suns energy connected with God'senergy, that is, that we have used two different energies through the beginning through which we have life, and that the tree of life, our spine, needs two special energies that sustain its life functions.

When we look at the picture of our spine, we see that it has 30 + 3 vertebrae, and that they are divided into the first 7; and these are the cervical vertebrae and is called C. (1, 2, 3, 4, 5, 6, 7).

The others are T and there are 12 of them.

The third is L and there are 5.

The fourth is Sacrum and there are 6.

The fifth is Colon and it has 3.

They all function in their appearance which is like a tree, from them grow nerves and channels through which life energy is transferred to our body and through which the nervous system has control over our entire body. Above all these vertebrae there is a large and small brain, and that means that over all these nerves, there is a brain that keeps under control our entire body in a physical state, that is; movement and work of all our organs. Likewise, above all creation, there is one who created all of this and through his wisdom and his power everything exists. He has control over all creation and he manages and directs the work of all creation. The small, brain controls our entire body and through the spine performs all the functions that can be performed in one healthy body.

The big brain, which is above the small, has control over our whole being that is the shape, spiritual and physical. That said, the cerebrum is also the center of all centers and it is a direct connection with God as if it were a direct connection with our whole body. Through the great brain and through it we have oneness with God, (through the spirit of God) which he breathed into us when he created us. Then God showed us, through the serpent and its perfect geometry the shapes that are seen on its body and that are also on us, that we are all one whole and that we were created by the same, Creator of the same elements that were in the Earth, i.e. the Earth was in the sea, in salt water, until the creation of life on it. This means that the Earth has elements of sea water in it and that it is composed of minerals in which it has been for thousands of years.

This means that water together with minerals is the main carrier of our life, the physical body, then the Sun is the center of the whole system of life on this Earth and without it and its light, which has power given to it by God on the 4th day of creation, to give life to the whole planet and to all that is seen and not seen. Thus all truths complement each other, showing the perfection of God's creation. This information shows: Seven vertebral located in the neck and which are right next to the head and brain.

The first nerve, which is also the first to the head, is in direct connection with both brains, as well as with the parts of the body that are in the head. And the very word head or the main says that this is the main center of all nerves.

The seven vertebrae also show the first seven days of the week which are the center of all the days that have been on this Earth and which will be in. As if they were the first days and were up to head to the Creator. Just like they are these first seven vertebrae first to the head on our spine. The first week is also the head of all weeks and on it all the information that time has carried for these 6.000 years. And where, in that first week, God showed us the way and the direction in which we need to go to come to the direction of our Creator Lord God. And at the same time, they show the seven days of the week that were the first days of creation. In the first week of creation, God created everything he needed to work and revealed all the secrets of life.

T - 12 are the nerves that keep alive all the vital organs which are called automatic organs because they function the way God left them to function, i.e. without our control, and on them lies the weight of our body carried by our spine. In that part of the body, just like the tree of life that carries the weight of the whole tree, so does our spine, in that part, those twelve vertebrae, also carries the weight of our body.

These twelve vertebrae, which are the main bearers of our body, is a pre-image of Jesus Christ who is on a tree that had such a shape of the letter "T" and which we need to open our eyes, that it was also the tree on which Jesus Christ died which is also called the cross. These nerves also have their name and shape and it is the shape of the letter TTTTe. - nerves, or vertebrae or nerves that carry a burden, life, or the body.

There are 5 vertebrae called L-5. They have their function in the spine to transfer the function of the lower part of the body with the upper part, and to cross the direction of their nerves in both halves in that part of the our spines, and to connect them with each other. These parts of the spine, which are located in the part below these T. 12 vertebrae, are also called the cross nerves because they are located in the part of the body called the cross. And their name says that through the cross or Christ a connection

will be established with two separate creations that will merge in their functioning and make the whole of life through the cross and the cross nerves. The nerves that have a direct connection with the Earth are connected in the crosses whose roots are in direct contact with the ground. Those nerves that are in its own special direction facing the Earth, and these are the nerves that are at our feet, they will be in the crosses and across the cross will be connected with the tree of life to the other vertebrae and together they will make the life of our body. As through Christ / the Messiah, who united the life that was created on Earth with life that gives life, and so united us with a tree of life found in Christ leading to eternal life.
Sacrum is the name of the nerves below the cross.

There are 6 of them and they have the function to move the part of the body of the lower extremities or the lower part of the body which includes the parts from the hips down and which serve for the movement of the legs and all the parts of the organs which are there, and for the function of the work of those parts of the body and organs.
The direction of these nerves go vertically, through our feet towards the ground the same as the three vertebrae and nerves together form the root which has contact with the Earth and which is, in fact, the last root of the tree of life of our physical being. At the root of the spine are also located reproductive organs, as well as organs that have their function to flush out toxins and waste from our body. In that part of the spine they

find the organs through which our body is cleansed, and like any waste, it returns to the Earth.

All these nerves work together to make the life of every creature, both human and animal, because animals were created on the fifth and sixth days of creation, just like Man, and they are dependent on their tree of life and on their backbone, as well as on the Creator who created them, in their form and who through this form showed that we are part of God's creation and that he planted a tree of life in them and spine. God says that he is the tree of life and that through him, and on him, everything there is that exists, that He has planted in us the tree of life that is a part of Him and thus we have become an integral part of His creation and perfection that is under His control, both people and the whole universe.

He showed us a picture of everything created through a snake, so that we could understand that we are all one whole of God and at the same time an integral part of the universe.

Adam / Adem lost eternal life and God has shown us that we are, in fact, created to use God's tree of life and to have eternal life through Him and through God. And that we can have it through the tree of life again, through Christ / the Messiah, who created all this, who showed us everything; in his words, letters, numbers, space, so we could all understand and keep it, because it is not difficult to keep it, with God's help.

The same way it showed us the information that was at that moment when through the serpent Satan - the Devil deceived the first man and the first woman, and when they, through their free will, refused obedience to God and their Creator, and when they lost their holy life, that at the same time they lost the connection through which they had the eternal life, i.e. the connection they used through the food that was in the tree of life, which was the food that God gave. It was in obedience to the law of God, and in which was the life energy that comes from God, because (He) is the life energy he uses and without which nothing created can exist.

When Adam / Adem lost the connection with the use of that food and that energy, at that moment the thread of life that gave him life was broken. By separating from it, death immediately began to enter his cells that did not have food from which they received eternal life. That is, at the moment when the direct connection with the source of life is broken, at that moment death begins to occur. It happened to the first man and the first woman as well as the snake. And the snake at that same moment, it lost its form, its body and its eternal life. The same thing happened to the first married couple, and they lost their sacred form because, at the same time they lost their sanctity, they have lost their sacred form, and eternal life.

At that same moment he is the originator of sin, (Satan - the devil), when he sinned and deceived man and for which everything began to die on this planet, both Humans and nature and animals, everything through Adam's / Adem's sin, he became a part of sin and lost eternal life. For everything, nature and animals as well as Man, all lived a holy life in which there was no death. But in that fateful moment, which happened, and in which (Satan - devil), thinking that he will deceive God and that he will set his throne above the stars of heaven, he is deceived because from that second, when he deceived man, God's creation which was the crown of God's creation, deceive himself, he got carried away by his beauty and his glory given to him by God, (that) he only uses it and through which he was supposed to glorify the Creator, he turned it against God, and wished that he transcribes it all down for himself, thinking that it was all his power and strength that did it. And in the moment he got carried away he did not remember that he is only a creation of God and that without God he cannot even move his body, and that God knows our thoughts and that he knew his too.

At that moment, he lost touch with the Creator and lost (the power) on which his life depended. From the same moment, together with Adam and Eve, he (began to die out) on this Earth. Because he also lost touch with the original energy and with the root of the tree of life through which he had eternal life, because he too was a created being of God who lived from the creator's power and energy.

This information says that time has expired for everyone, and that time will stop at 6.000 years or 9.000 years and that this form of life will end, that Omega reached Alfa and in that union disappeared the one who conceived sin as well as the sinner.

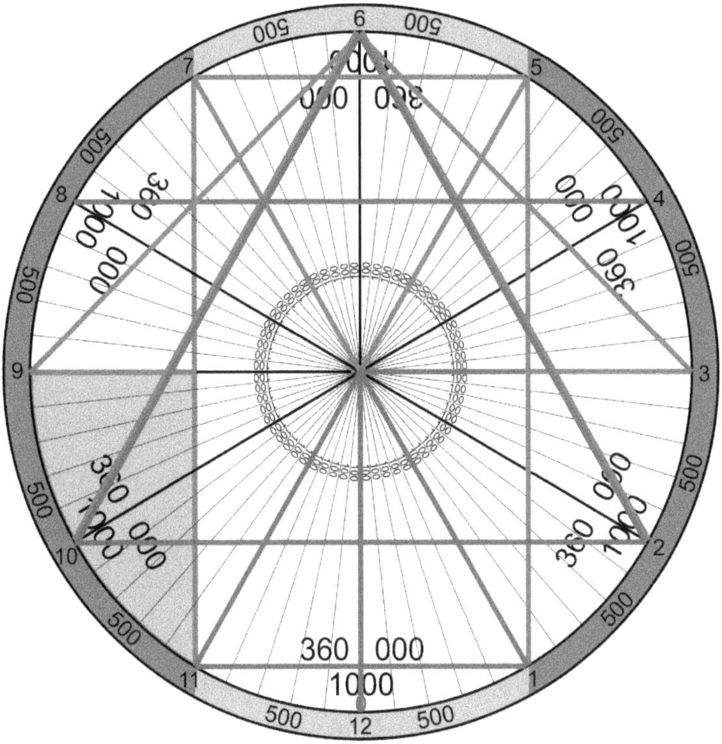

Picture 53 (ALPHA AND OMEGA IN TIME AND SPACE)

It shows a 360 ° circle containing all the Alphas and Omegas that were Alphas and Omegas and, by God's providence, sustained life of everything created on this Earth and in this Galaxy. This picture shows that everything, in the only great whole that has its limit of becoming and in that limit of time and space and through numbers and letters the Lord God has revealed to us all his secrets, in which he created life and in which he will end life on the same foundations on which he began.

For Jesus / Isaa / Christ / Messiah says "I am the Alpha and the Omega, I am the first and the last and I am the beginning and the end." Here you can exactly see the hours, days, years and thousands of years which show 6.000 years. And those 6.000 years are in this border or Omega in which they existed. You can also see all the letters that make up Alpha and Omega. All the numbers and all the roots of the numbers are visible. You can see the cross from Golgotha, and only one person died on that cross, and that person shows and proves everything visible and invisible, and he says that this person is the center of the Universe because through him everything that has become and everything that will become. Because through that person, who will raise us from the dead and through whom we will get our life back, and that very person was on the cross, where he gave his life for us. On this cross a replacement of life for life and death for death was created. Only those who understood that Christ is the Messiah through His death, who by the will of His father and our father, came to this Earth to give His life, for everyone who through His life, seeks eternal life from God. And only for the people of this world because they have been deceived by a strong opponent (Satan the devil) and a liar, who is the father of lies because he told the first lie and deceived one third of the Angels of God. But Jesus did not give His life for them because they had 3.000 years' time of grace, to repent and seek forgiveness from God who created them. They had exactly as much time as Adam/ Adem lived in the Garden of Eden until he deceived him.

The deception of people and Angels arose from the envy that happened in him. The Creator created him to work, exactly gave to him a specific function, and had the highest position in Heaven right now, after Jesus / Jesus / Christ.

He was the shield of God's power and light. He was Cherubim, special Angel, his name was in Heaven, Lucifer, which means light bearer, because the word Luc itself means light, and the word fair means fire or light again. He got that name because from the nearness in which he was, while he was blocking the light of God and he himself had the light on him.

Explanation: When Moses / Musa was on Mount Sinai, when the Lord God gave him his law, he could not look into the light of God and he had to cover his face. So did the children of Israel, they covered their faces from that strong light, even though they were at the foot of the mountain. And God says that no one can see the face of God and stay alive for there is so much power and energy in God that neither Angels nor Men cannot bear it. Only the Son of God can see God. So all creatures created are not able to see God and stay alive.

Likewise, that fallen Angel could not see God and stay alive. But he was nearby and from it he got the power that made him arrogant, thinking that it was his and that he can have that power the way Jesus / Isaa had. Jesus is special, he is the first created (Son of God) to whom his father gave his strength and power and handed everything over to him and who has one will with his father and who does only the will of his father. That Angel thought that it was also his light, which shines similarly of Christ, that it is eternal.

But he was created by Jesus / Isaa, as well as everything that is seen and not seen because the father gave him the hands to do the will of his father and only Jesus / Isaa, can approach his father and stay alive, because God gave him of his strength and of his own power, power into his hands and to do one will with his father.

So Jesus is not an Angel, he is also the first created (Son of God) created by the Heavenly Father.

Because he is the only created Son, created by the Heavenly Father, everything else was created by Jesus, because the Father gave him the strength and power to create everything because he is one with the Father and has one will. When Jesus Isaa was on Earth, he said to the Pharisees, “Me and my father we have one will, and whoever sees me has seen the Father, because I did not come into the world of my own free will, but the will of my father.”

This information says that Satan - the devil - had the same meaning as the Sun in the first week of creation when God created it on the 4th day of creation and gave him the strength he needed to sustain life on this planet. He also gave him the light and strength that was required to perform his duties. And everything created is the creation of God, and the throne of God is in (Heaven), and which can at any time interrupt life in every creation that he created, because all creations have a reason why they exist.

At that moment, when (Satan-devil) deceived Adam / Adem and when he tried to steal for himself the throne - of God, lying to Angels and Men, in order to have as many supporters as possible he used for his own purposes. He tried to make his throne on this Earth where he would be respected and listened to by deceived Angels as well deceived people. The throne on which he would be the God of this planet. And that people live in these conditions in which they still live. Because he does not have the strength to give people and fallen Angels anything that would give them eternal life, because he does not have it either.

Everything he had and what he currently has is based on lying or stealing because he tried to steal the throne of God with his lies and theft. But the very fact that he separated himself from eternal truth, eternal life, eternal light, and all this of his own free will, led him to lose everything and become the one who, along with his sins, will die. Heavenly Father has made a plan of salvation for every soul, so that through his second chance he can return to eternal life. Every soul has the same chance and no one will be to blame if a person, for his personal reasons and prejudices, does not use that chance. And the serpent became enraged with the woman, and went to fight with the other seed which held the commandments of God and had the testimony of Jesus Christ.

"Revelation of John"

CHAPTER 14:

Verse 1. "And I saw, and behold, a lamb standing on Mount Zion, and with it 144.000 who have his father's name written on their foreheads."

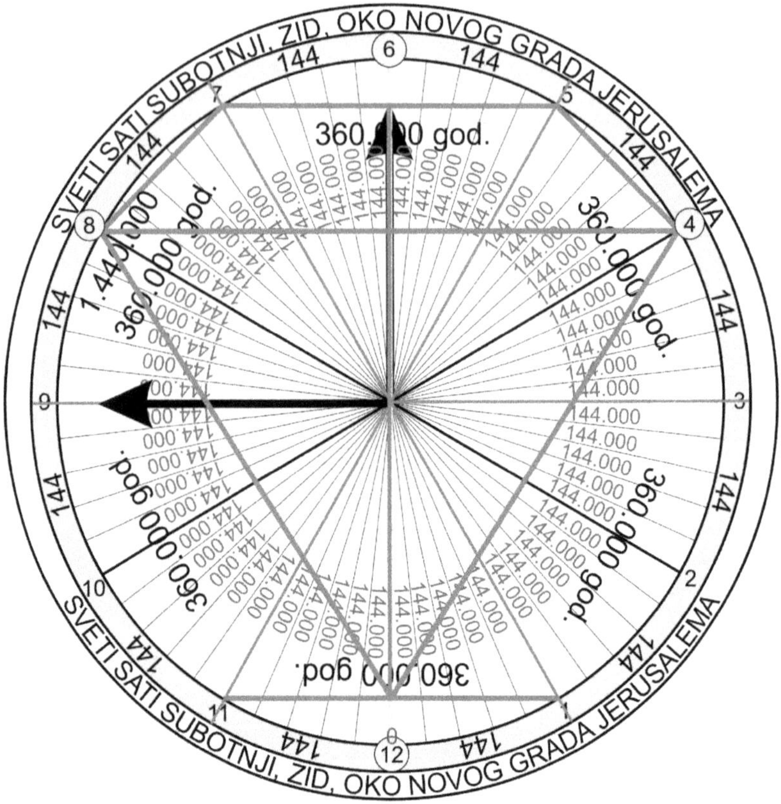

Picture 54 (TIME THROUGH TIME IN HIS TIME).

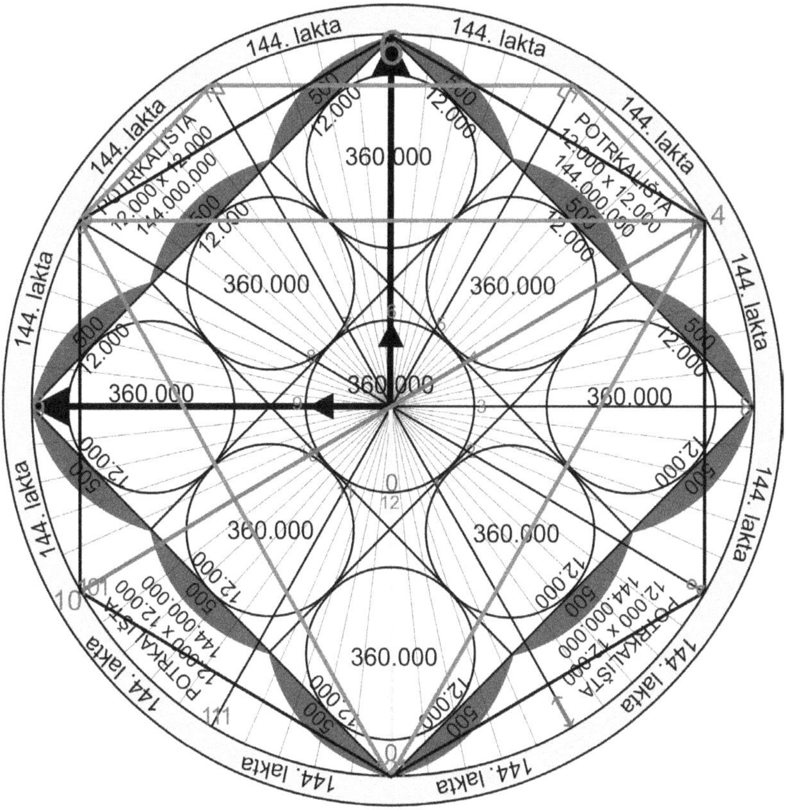

Picture 54a (TIME THROUGH TIME IN A NEW FORM OF TIME).

This number that the Creator showed us, through his word, will 144.000 be in the new city of Jerusalem with the lamb - Jesus / Isaa, and that they will be his followers, this is also the number that leads us into the first week of creating life on Earth.

On the sixth day of creation, in which God finished creation and to which he created all that he created, and the last he made Man as a (crown of creation). On that day, on Friday, which is also the sixth day of creation, there was also the 144th hour of creation, which means that man was created in the 144th hour of the beginning of life on this Earth.

This number 144 appeared immediately in the first week of life and on the first day that man began to live, which was Friday, 144th hour since the beginning of life on this planet.
This is also described in the first page of the first book, Moses-Musa, which was the beginning of the description of all the truths that God showed to the people through whom he served, to transfer through them his information, necessary for an easier understanding of the truth.
About the beginning of the life of all creation, who created it and who maintains it in life all these 6.000 years; Ever since they started to convey information orally, through God's chosen people, and were later transferred to various papers that were then at the time. They are constantly being rewritten on new papers that are perfected in their quality to have a longer lifespan. Every guiding book shows that the Universe has its truth which confirms the truth through true and truthful events.
Only the truth could preserve the truth, which confirmed itself through space and time.
Thus, this number 144 also has a truth that confirms itself, and that is:

1 week has 144 working hours:

10 WEEKS has 1.440 working hours:

100 WEEKS has 14.000 working hours:

1000 WEEKS has 144.000 working hours:

Or 144.000 thousand, who the word of God says are special, are counted separately, these are people who have suffered great hardships because of their beliefs in Jesus Christ and who will be with Christ in the new city of Jerusalem.

Related to this information:

Every day of the year bears its own fruit just like the tree of life in the midst of the new city of Jerusalem that bears its fruit.

Like the tree of life that is in the middle of the Garden of Eden, which has given and will give, again its fruit; for he writes, in the word of God; that the tree of life is in the midst of (the new city of Jerusalem), and that it bears its fruit every month; Likewise our time in which we live shows; that every day and every month, it bears its own fruit.

1 day = (400) people, just like the new city of Jerusalem. There is the shape of a cube and which at (4) the corners, stands the length, width and height the same.

10 days = 4000 people.

30 days = 12.000 people, which is a number for 1 month

Just as the tree of life will bear fruit every month, so it will give its fruit on this Earth every month, as it was giving. If it gives 12.000 every month, then one year has 12 months and multiplied; gives this result 12.000 x 12 = bears the fruit of 144. 000. This information says that every year there was a number of 144.000, which has been transmitted throughout the life of this Earth, and the tree of life gave its fruit every month.

And that is:

1 year has 144.000 thousand.

10 years has 1.440.000.

100 years has 14.400.000.

1000 years has 144.000.000.

This means that every 1.000 years beard its fruit of 144.000.000 thousand, this result comes when we multiply; 12.000 x 12.000 = 144.000.000, which is a time period of 1.000 years.

And that's exactly 144.000.000 x 6.000 = 864.000.000 in 6.000 years, this means that this number fits exactly into the perfect number which show us the accuracy of all perfections. Because when on this number of 864.000.000 we add from 36.000.000 = 900.000.000, that is a number that shows the perfection of God which shows and reveals itself through his information. In this number, of 36.000.000, which shows that in this number it is located the years in which the consecration of the land of 1.000 years will take place, on this number are all the numbers of seconds, minutes, hours, days, years, or numbers of people who will be saved from this world, but it is (separated), from the time by which is counted 6.000 years.

In whichever direction we start adding seconds, hours, days, years and thousands of years from 0 to 6.000 or 9.000, we get the same perfection of number, hours, days, years same as space, we get the number (360) and in it we find the size of space, the length of time, as well as the sum of all numbers, which have their meaning and through their (code) which reveals; the secret of these dimensions, and the message contained in them. Where it proves and confirms everything, that it fits to the millimeter each of these creations of God into each other.

This information is calculated by the hour of time, by the number of days and years in these 6.000 + 3.000 = 9.000 which passed + 1.000 a year that will come only when life on Earth ends in this form in which we now live. It's also called 1.000 years of consecration of the land. Years from 0 to 6.000 years were counted on the same time line. The same counting goes through seconds, minutes, hours and days, all in their numbers through the number 3 + 6 or 360. It shows the common whole on which the Lord God, in his measure, set space as 360 °, time as 360 seconds or minutes, days and years.

Such as numbers; 3 and 6, make a number 9, in which it showed (Code and direction). It also showed life in the code. That

everyone who asks has a confirmation and in that confirmation finds information about who created him, redeemed him, and who will recreate him, before whom the book of his life will be opened in which he will see the deeds and faith of each individual, on the basis of which he will be weighed, according to the standards of God. And he revealed to us the time, which passed this time, as well as to the complete creation, which God created on this planet and to men, showed that, from the counting of time beginning from the first second of life on it, time is counted in; seconds, hours, days, years and thousands of years; And they said the same thing, 6, 60, 6.00, 6.000 or 144 hours, in which the hours of the first 6 days of creation were shown, or number 6 showing time, from the first 60 seconds to 6.000 years:
or 360 days from the first year,
or 360.000 days of the first 1000 years,
or 3.600.000 days of 10.000 years, where it ceases to count the time and space.

But one important piece of information says that all these zeros, which are added to these numbers are just confirmations of the time in which they happened so that people would have confirmation of these events and of the time in to which they have happened or in which they will happen. And that means that the roots of all these information's are in the numbers 3,

6, 9 = or 3 + 6 = 9; or 1 + 4 + 4 = 9; or 0, 1, 2, 3, 4, 5, 6, 7, 8, 9, through these numbers that are the roots of all numbers we get this information that God gave us to know the truth that leads to eternal life.

And important information says that the number 144.000 = the number which shows the perfection of God's plan of salvation in which he revealed himself to us and it has been shown in all possible directions that he is the only God and that he created everything, and this number 144.000 still says:

When we break this number down into its parts we will have this information:

First, it is pronounced (100); and that is one whole on which is fulfilled ¼ time and space because 1 year has 360 days and 100 years has 36.000 days.

And 36.000 days divided into 4 parts are 36.000 : 4 = 9000 days.

Or 36.000 days times 4 parts is 36.000 x 4 = 144.000 days.

That says that every 100 years there are 144.000 days in it. Then after 100 follows the number (40); when we disassemble it, we hear and see that it says 4 x 10 or 40 and these are 4 time periods that have happened on this earth and that have a direction to take us to a time that we need to understand what God wants to tell us, and that is: 4 events happened on this Earth in which God has shown us.

For the first time since the beginning of life on this Earth, at a time when the flood happened and when a whole generation of people disappeared in the flood and it is written: "And on that day all the fountains of the great deep were destroyed, and the constitutions of heaven were opened. And it rained on the ground 40 days and 40 nights."

Here the information was shown that in the first, 2.000. years of life on this Earth, just as God created water on the second day of creation, i.e. that he separated water from water and that he created living water which comes from Heaven and sustains the life of all creation through the power of God. And that it was in those 2.000 years, he washed the Earth with water from the sins that took place on Earth and because of which the Creator destroyed the Earth and the people who produced those sins.

Only 8 souls remained and they carried life through the flood, and from which all this generation was born. Here it is seen that in that flood the earth was washed away from sin and the water fell from the Earth and from the Sky for exactly 40 days.

And the following explanation says that Jesus / Isaa, before the beginning of his mission, because of which he came to this Earth, i.e. Before he started with his work in which the power and love of God was revealed, when he began to preach the Gospel and to show the fulfillment of the word of God, so people understood that he was the one spoken of by all the prophets, for whom all time and space said, that he is the one, that they might believe and that through faith, they may understand the meaning of all rituals, and that he is a lamb of God who was slaughtered and through whom sins were forgiven through faith and were promised, and that the time has come for that lamb of God to give his life, just as it has been symbolically shown for 4.000 years.

And before Jesus / Isaa ,began his mission, he went to the Jordan River, where a man was born to immerse him in the river in which he too, though sinless, was born by birth to a woman who had inherited sins from Adam / Adem, and he also washed his inherited sins, and left for us instruction that water, at the same time is a water grave in which the sins of all those who understand that they are sinners and that just by good deeds they cannot rovide themselves with eternal life.

That cleansed of sins, which once buried in a watery grave, arise clean, and through faith in God, keeping the law of God and through Christ, who is the tree of life and who has life in him, that through his life they may live.

In the flood and through the water he showed the first 40 days or 4 x10

The second number 40 shows, when the Israelites went through the desert towards the promised land, they went for 40 years.

In this information we see that behind the water came the desert and the drought through which the people had to go, through which everyone who had to go through who is written in the book of life. For every individual, before he comes to the knowledge of truth and Gods blessings, we must first go through the water of purification, as well as through the desert of thirst and heat that we need and which, in its own way, cleanses the soul and body. Likewise, Christ / Messiah went through both cleansings and showed us what everyone who wants his name, written in the book of eternal life will go through it too.

The third number 40 was shown when Jesus / Isaa, after purification through which he passed through the water grave, immediately after that he went to desert and fasted for 40 days without bread and water. And in that solitude he prayed to his father to give him the strength to endure the hardships he had to go through for the next 3.5 years, for as long as his Messianic work and to give him the strength to complete the work for which he came to this world, i.e. to fulfill the plan of salvation, to overcome death by his death, and to give life by his life.

The first thing he did, after the watery grave he left, through his mother, inherited sins, is that he fasted for 40 days. In that fasting, he overcame the appetite that made Adam / Adem lose eternal life. And he showed people that through the appetite there are many problems related to eternal life as well as to life on this world. Through appetite, many lives were extinguished prematurely and because of it many lives will lose eternal life.

Fourth number 40 says that Jesus / Isaa, after his death, resided on this world in human form, before he returned to his father and to our Father, exactly 40 days.

It was his holy; spiritual and physical form in which he put on after death and the completion of the fulfillment of the plan of salvation. That form he will wear for ever and ever, as a memorial of his sacrifice and his own love, as a memorial of his Father's sacrifice and the love they endured, for the people of this planet and for the disobedience shown by the first created man Adam / Adem according to the law of God. This information showed that the number 40 has been shown 4 times so far, which is also 4 X 10.

These are also the four corners that the new city of Jerusalem has.

So far it has proven, 100 as a whole or 100% a whole of everything or truth.

Then, we have 4 x 10, four sides in a cube that is located as a whole of 6.000 years or 9.000 years, because only the word 144.000 says that that word forty has 4 x 10 and that is 4 times, which refers to the 4 important events that Jesus / Isaa showed, how they happened to him that it will happen to some of his followers and that all who want to follow his path, which is the path of truth and life.

Before we reach the goal, we need to go through water, heat and hunger so that, only after death, we have the life he went to. Let's not get too attached to this life, because we are just passing by through it and that through this life, eternal life is conditioned. And that we are not just passing by on this Earth because through this life we will be, weighed for eternal life.

But the number 4.000 remained and he is the whole of 144.000. These 4.000 years, they say, that is true, for 4.000 years of life on this Earth, the plan of salvation was fulfilled. That our creator, our redeemer and our savior Jesus / Isaa / Messiah, about whom this information speaks about, where they complement each other through time, space and numbers and which indicate only one event that has occurred in this galaxy of ours and on this planet of ours. All other events they lose their significance and are not important at all because fateful events took place for our lives.

And on, the number (100) + (4 x 10) + (4000) = 144.000, which was repeated in every hundred, be it seconds, minutes, hours, days, years, or thousands of years. In numbers, and in letters, everything is confirmed , and proves who created us, the name is the Lord God and who will create us again. One to eternal life, and one to eternal death. Depending on our free will that we had in this life in which we lived and through which we by our free will ruled our desires and our feelings. Each number 100, when divided into 4 parts or ¼ = 25. And it is also the 4th part of 1 whole, regardless of its size, because each whole is also 1 one 100%.

Each whole, which makes a whole of 10, 100, 1.000, 10.000, 100.000, 1.000.000, 10.000.000, 100.000.000, 1.000.000.000, 10.000.000.000, zero can be infinity. All of them say that the first unit starts from number 1 and that this number, which turned out to be 12.000 x 12.000 racecourse that the new city Jerusalem has and its length is as wide as its width. And that means 12.000 x 12.000 = 144.000.000, as it is said in the holy word of God. 12 is the number or root on which the first day arose and by which God began to count the first day, and through time we have entered the dimension of time which is proportional to the time in which we have spent it in this life.

It starts with:

1 year has 12 months.

10 years has 120 months.

100 years has 1.200 months.

1000 years has 12.000 months.

When these 12.000 months are multiplied by the number of people; which gives the tree of life every month, because it is the number of people whose names are written in the book of life in (Heaven) and comes from the sum:

Every day 400 people, for 10 days 4000 for 30 days is 30 x 4000 = 12.000. When we multiply that sum by the months in 1.000 years, it is 12.000 x 12.000 = 144.000.000.
From this result we get the sum of people who enter every 1.000 years, when we multiply them by 6.000 years we get 144.000.000 x 6000 = 864.000.000.
Then, 6.000 years, when we multiply by years and days it is; every 1.000 years there are, (360.000) days x 6; or 6.000 years.
We get the time or space of the new city of Jerusalem, and that is 360.000 x 6 = 2.160.000 days, plus 3.000 years of sacred time from the Garden of Eden = 1.080.000 days + time of consecration of the land = 1.000 years or 360.000 days.
All together it is:
2.160.000 + 1.080.000 + 360.000 = 3.600.000 days.

This information says that the new city of Jerusalem, through the time that was counted on this Earth, and through this time will be counted, the people who were born on this Earth, who will pass through the 12 gates, which are located on (4) four sides; The new city of Jerusalem, because it stands on four corners. And on each side (3) three doors.
Which unlocks Jesus / Isaa / Messiah and through which people will enter through Christ's merit. That time, which was counted through the days and months, which are added up on this Earth and which fit into the exact days that have passed, those days,

that time and that space are a measure of space and the number of days and times by which God measured the size of the new the city of Jerusalem. On a new world and in a new life that will be and a holy life, lost by our forefather and restored by our savior, Jesus Christ.

And it counts: The city stands on four corners, that means it's a cube, and that all its sides are equal in height in width and in length. It shows time that counts in a precisely determined goal, and that is, when time is added in which it stands, it is the time of one day or time of one year; or the time of one day divided in 4 parts, which is: 12 hours: 4 hours = 3 hours. There you get 3 doors.

When we divide 12 months into 4 months we get 3 months; 3 x 4 = 12. This shows us that the hours, months and years are also divided in time and to point to the 3 gates located on the 4 sides of the Earth or cubes, as described by the word of God, that the new city of Jerusalem stands at 4 corners.

Each side also has three doors through which people will enter the eternal life, Garden of Eden. And that, exactly, in a number that showed time and the Universe. And it says:

Through your own direction, through which you need to show yourself exactly through time, all that has so far been shown and revealed, has been revealed through space and through time.

Since the number 4 has its own code and goes through time, then this is the order of that direction or truth that needs to confirm

itself, because everything what does not confirm itself is not true either.
Number 4 says, one day has 12 hours, divided into three doors, which are in the new city of Jerusalem: 12: 4 = 3

Verse 2. "And I heard a voice from heaven, as the voice of many waters, and as the voice of a great thunder and I heard the voice of harpers harping with their harps."
Verse 3. "And they sung as it were a new song before the throne, and before the four beasts, and the elders: and no man could learn that song but the 144.000, which were redeemed from the Earth."
Verse 5. "And in their mouth was found no guile: for they are without fault before the throne of God."
This information speaks for itself; that these 144.000 are the number of those who stand without blemish, before the throne of God.
And without flaws, there are no people on this Earth.
The Word of God says that only one person was on this Earth without flaws. He was born, lived and died without any defect. This is Jesus Christ / Messiah.
But even these people before the throne are flawless.
Only those who have been redeemed by the blood of the Lamb, the Christ / Messiah, can be without blemish. And if they are not cleansed in the blood of Christ, they cannot even enter the new

city of Jerusalem. That is to say, that all who come again to eternal life or in the Garden of Eden, or the new city of Jerusalem, will be without blemish and will be cleansed with the blood of the Lamb slain since the existence of the world.

This information says, all people who find themselves in the eternal life again, and there will be as many of them as perfection says in its perfect number and its perfect whole that passes through space time and number, as written in Revelation:

CHAPTER 14:

Verse 6. “And I saw another Angel fly in the midst of heaven, having everlasting gospel to preach unto them that dwell on the Earth, and to every nation, and kindred, and tongue, and people.”

Verse 7. “Saying with a loud voice, Fear God, and give glory to him; for the hour of his judgment is come: and worship him that made Heaven, and Earth, and the Sea, and the fountains of waters.”

Verse 12. “Here is the patience of the saints: here are they that keep the commandments of God, and the faith of Jesus.”

Verse 13. “And I heard a voice from heaven saying unto me, Write, Blessed are the dead which die in the Lord from henceforth: Yea, seith the Spirit, that they may rest from their labours; and their works do follow them.”

Verse 14. “And I looked, and behold a white cloud, and upon the cloud one sat like unto the Son of Man, having on his head a golden crown, and in his hand a sharp sickle.”

CHAPTER 15:

Verse 1. “And I saw another sign in Heaven, great and marvellous, seven Angels Having the seven last plagues; for in them is filled up the wrath of God.”

Verse 18. “And there were voices, and thunders, and lightings; and there was a great earthquake, such as was not since men were upon the Earth, so mighty an earthquake, and so great.”

Verse 20. “And every island fled away, and the mountains were not found.”

Verse 21. “And there fell upon Men a great hail out of Heaven, every stone about the weight of a talent: and Men blasphemed God because of the plague of the hail; for the plague thereof was exceeding great.”

From these verses we see one part of the last moments of life, on this world. The moment when God removes the Sun and the moment when there will be a polar winter, as it was before the beginning of life on it. When the last moment of its time strikes it, which was in this form on which we now live.

Lord God, showed us through the letters of the alphabet, which at the root there are 24, the same as in one day and one night there are together 24 hours and that fit together and show their information coming from that direction, and that is that every letter and every number show their information. The letters of the alphabet have the shape of a gene that show each letter, just as that letter looks, and it also shows through the number, where that gene serves and where it is located, in to which part of the body and to which organ it belongs. And on which spine the vertebra connects our organs with the tree of life.

All letters have a root, from the letter A to the letter Z, it is also 24 words.

All numbers have a root from number 1 to number 9.

It has a root of 0 to 12 all the time.

All space has a root from 3 to 6.

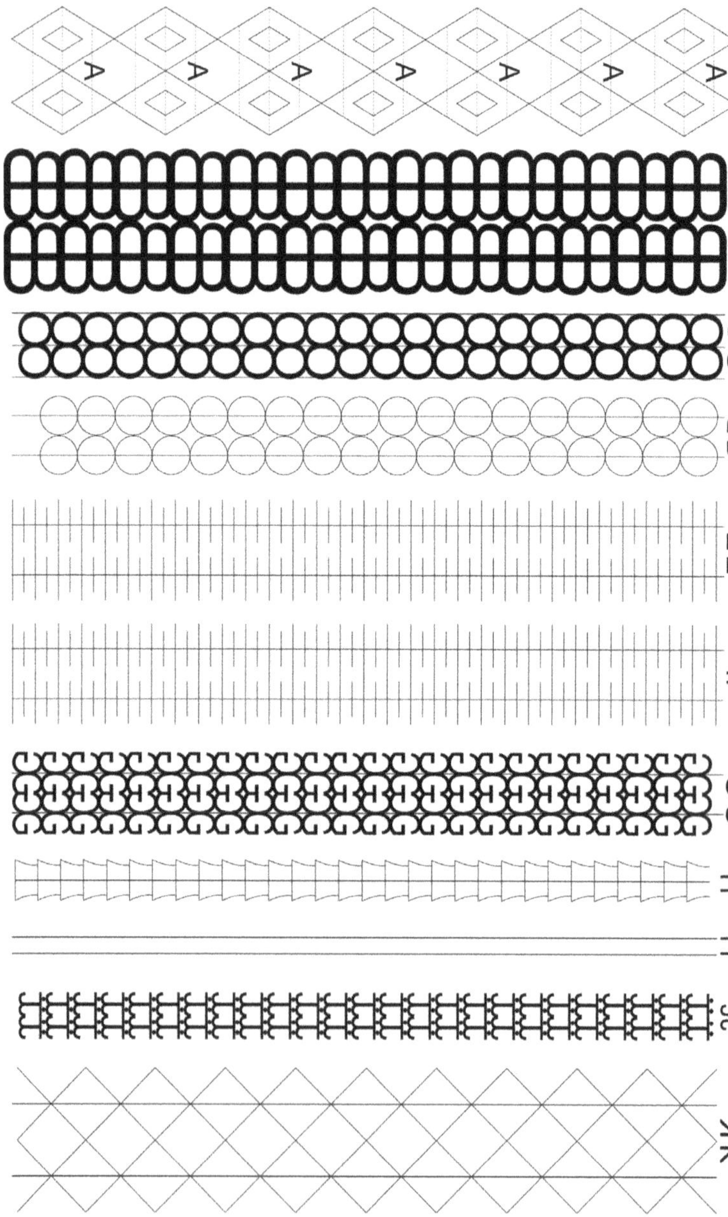

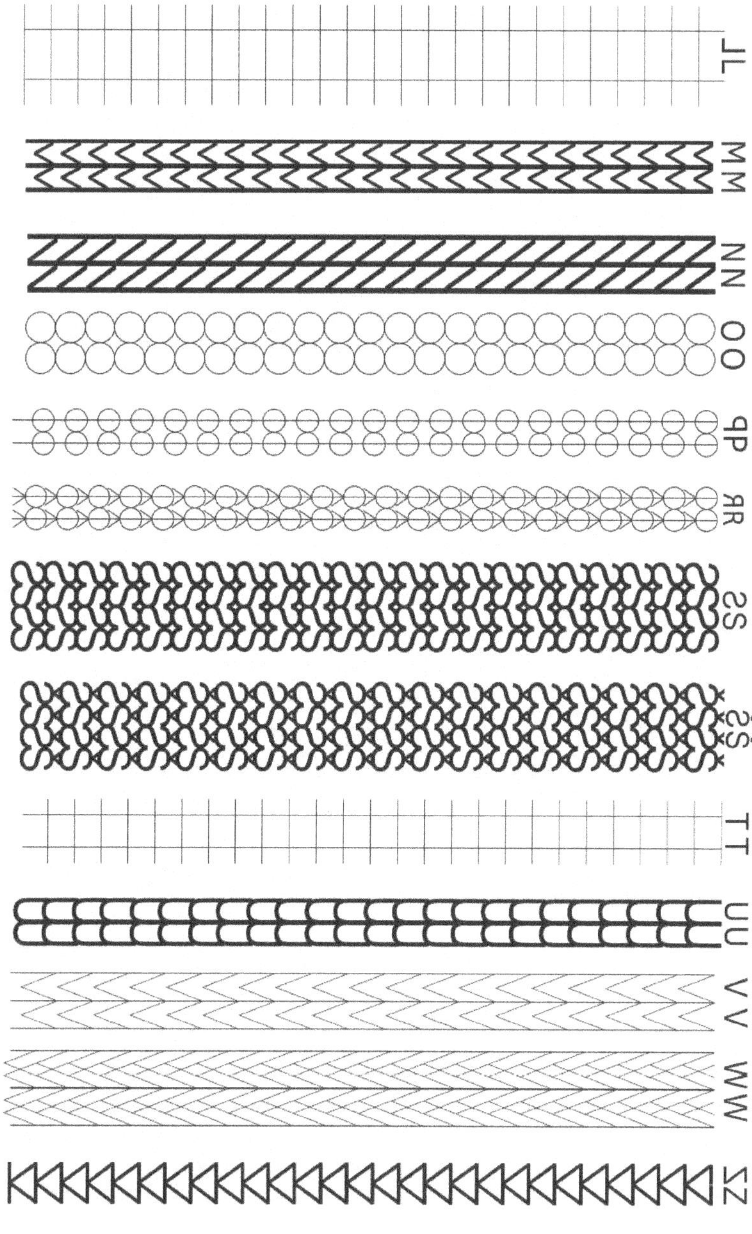

Picture 55 (LIFE THROUGH GENE AND LETTERS).

It shows the form of the genes classified according to its Universal appearance, what forms of genes exist in our galaxy and in us, who are God's creations and an integral part of the Universe.
It shows and indicates how the universe is multiplied, i.e. space in the universe and life in it. And that this world is a copy of life in the Universe, just as Man is a copy of the Universe just smaller. God showed us all this, so that we would not look for other Gods and invent them ourselves, and prayed to them.
Man is a creation of God, who is also a copy of his whole life in the Universe. As people multiply, so the Universe multiplies in its existence and in its form.
The universe, just like man, depends on the Creator and is connected to its Creator. And without the Creator God cannot exist, because only God's power and God's energy has the power to keep it all alive and to renew it all, and to multiply it in that particular creation, for which it was created.

CHAPTER 21:
Verse 5. "And he that sat upon the throne said, Behold, I make all things new. And he said unto me, Write: for these words are true and faithful."
Verse 6. "And he said unto me, It is done. I am Alpha and Omega, the beginning and the end. I will give unto him that is athirst of the fountain of the water of life freely."
Verse 7. "He that overcomes shall inherit all things; and I will be his God, and he shall be my son."

Verse 8. "But the fearful, and unbelieving, and the abominable, and murderers, and whoremongers, and sorcerers, and idolaters, and all liars, shall have their part in the lake which burns with fire and brimstone: which is the second death.

CHAPTER 22:

Verse 6. "And he saith unto me, These sayings are faithful and true: and the Lord God of the holy prophets sent his angel to shew unto his servants the things which must shortly be done."

Verse 7. "Behold, I come quickly: blessed is he that keeps the sayings of the prophecy of this book."

Verse 10. "And he saith unto me, Seal not the sayings of the prophecy of this book: for the time is at hand."

Verse 11. "He that is unjust, let him be unjust still: and he which is filthy, let him be filthy still: and he that is righteous, let him be righteous still: and he that is holy, let him be holy still."

Verse 12. "And, behold, I come quickly; and my reward is with me, to give every Man according as his work shall be."

Verse 13. "I am the Alpha and the Omega, the beginning and the end, the first and the last."

Verse 14. "Blessed are they that do his commandments, that they may have right to the tree of life, and may enter in through the gates into the city."

Verse 16. "I Jesus have sent mine Angel to testify unto you these things in the churches. I am the root and the offspring of David, and the bright and morning star Danica."

Verse 18. "For I testify unto every Man that heareth the words of the prophecy of this book, If any man shall add unto these things, God shall add unto him the plagues that are written in this book."
Verse 19. "And if any man shall take away from the words of the book of this prophecy, God shall take away his part out of the book of life, and out of the holy city, and from the things which are written in this book."
Verse 20. "He which testifieth these things saith, Surely I come quickly. Amen. Even so, come, Lord Jesus."
Verse 21. "The grace of the Lord Jesus Christ be with us all. Amen."

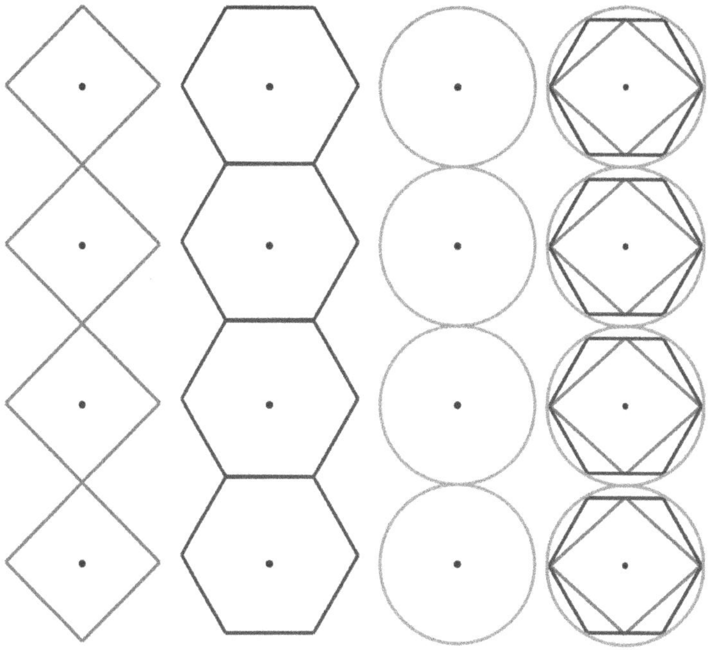

Picture 56 (WAYS OF THE POWER AND ENERGY OF GOD - THROUGHOUT LIFE UNIVERSE).

In these pictures, you can see the directions and lines through which we are connected with the Creator; of life and the Creator of the Universe. Through which all created takes its life (energy) and through which, the Creator maintains the life of everything (created). These are just; basic geometric forms, on which all creation exists and lives. Through the power of God and his life's energy. Which goes its own way. From the creator of life (and) who went around his circle of Alpha and Omega; and returned again to the one from whom it originated.

And he created all creation, the Creator whose name is the Lord God who has his throne in heaven, and who by his power and his (wisdom) creates and maintains life in his creations, and who shows through all creation: That only he is the Creator and God and that there are no other Gods besides him. And that everything belongs to him alone (glory and thanks). And that we will all come (only) before him and receive the salary we have earned of our own free will in this life.

AMEN.

CONTENT

www.ingramcontent.com/pod-product-compliance
Lightning Source LLC
LaVergne TN
LVHW050908210926
848482LV00012B/374
* 9 7 8 1 9 2 2 8 3 9 6 6 4 *